AF380650

Optofluidic Devices and Applications

Optofluidic Devices and Applications

Editors

Francisco Yubero
Fernando Lahoz

MDPI • Basel • Beijing • Wuhan • Barcelona • Belgrade • Manchester • Tokyo • Cluj • Tianjin

Editors
Francisco Yubero
Instituto de Ciencia de Materiales de Sevilla (CSIC–Univ. Sevilla)
Spain

Fernando Lahoz
Universidad de La Laguna, Santa Cruz de Tenerife
Spain

Editorial Office
MDPI
St. Alban-Anlage 66
4052 Basel, Switzerland

This is a reprint of articles from the Special Issue published online in the open access journal *Micromachines* (ISSN 2072-666X) (available at: https://www.mdpi.com/journal/micromachines/special_issues/optofluidic_devices_applications).

For citation purposes, cite each article independently as indicated on the article page online and as indicated below:

LastName, A.A.; LastName, B.B.; LastName, C.C. Article Title. *Journal Name* **Year**, *Volume Number*, Page Range.

ISBN 978-3-03943-717-7 (Hbk)
ISBN 978-3-03943-718-4 (PDF)

Contents

About the Editors

Francisco Yubero is a research scientist from Consejo Superior de Investigaciones Científicas at the Institute of Materials Science of Seville, Spain. He received his Ph.D. degree in physics at the Universidad Autónoma de Madrid, Spain, in 1993. From 1993 to 1996, he worked as a Postdoctoral Fellow at the physics department of the University of Odense (Denmark) and the Laboratoire pour l'Utilisation du Rayonnement Electromagnétique (France). His work focuses on the development of theoretical models to simulate electron energy losses of swift electrons traveling near surfaces; thin-film oxide synthesis with tailored properties, through the control of their microstructure and chemistry, to achieve specific photonic, electric, electrochromic, luminescent, or electrochemical functionalities; and the development of optofluidic transducers based on microstructured multilayers as local temperature sensors based on Fabry–Perot resonators or liquid sensors based on porous Bragg microcavities.

Fernando Lahoz is a Professor at the Physics Department of the University of La Laguna, Spain. Before joining the University of La Laguna, he was a Postdoctoral Research Fellow at the CNRS Laboratoire d'Analyse et 'Architecture des Systèmes in Toulouse, France, for two years, working on optically active waveguides, lasers, and optical amplifiers. He received his Ph.D. degree in Condensed Matter Physics at the Universidad de Zaragoza, Spain, in 1996. His current research interests include optical properties of glasses, glass-ceramics, and nanocrystals for up-conversion processes, photovoltaic applications, displays, and optical sensors. He is interested in fluorescent proteins and fluorescence in biological systems as well. His work also focuses on the development and study of optofluidic devices for high-sensitivity optical sensors, portable optofluidic lasers, and optical resonances in microfluidic systems.

Editorial

Editorial for the Special Issue on Optofluidic Devices and Applications

Francisco Yubero [1],* and Fernando Lahoz [2],*

[1] Instituto de Ciencia de Materiales de Sevilla (CSIC–University Sevilla), 41092 Sevilla, Spain
[2] Departamento de Física, IUdEA, Universidad de La Laguna, 38200 Santa Cruz de Tenerife, Spain
* Correspondence: yubero@icmse.csic.es (F.Y.); flahoz@ull.es (F.L.)

Received: 9 September 2020; Accepted: 20 September 2020; Published: 23 September 2020

Optofluidic devices are of high scientific and industrial interest in chemistry, biology, material science, pharmacy, and medicine. In recent years, they have experienced a strong development because of impressive achievements in the synergistic combination of photonics and micro- and nano-fluidics. Thus, sensing and/or lasing platforms showing unprecedented sensitivities in extremely small analyte volumes, and allowing real-time analysis within a lab-on-a-chip approach, have been developed. They are based on the interaction of fluids with evanescent waves induced at the surface of metallic or photonic structures, on the implementation of microcavities to induce optical resonances in the fluid medium or on other interactions of the microfluidic systems with light. In this context, a large variety of optofluidic devices have emerged, covering topics such as cells manipulation, microfabrication, water purification, energy production, catalytic reactions, microparticle sorting, micro-imaging, or bio-sensing. Moreover, integration of these optofluidic devices in larger electro-optic platforms represents a highly valuable improvement towards advanced applications, such as those based on surface plasmon resonances, already in the market.

In this Special Issue on Optofluidic Devices and Applications, we include 10 papers, covering different aspects related to water quality monitoring (1) and purification (4), structural stability of nanohole array-based devices (2), hydrodynamical focusing (3), fluorescence-based thermometry (5), electrofluidics (6), electrowetting (7) and optofluidic manipulation (8). Additionally, there are two interesting review papers on opto-electrokinetic-based manipulation and fabrication (9) and portable optofluidic systems for ocean monitoring (10).

In particular, Zuo et al. describe an optofluidic device for precise real-time detection of dissolved oxygen based on the plasmon resonant shift of silver nanoprisms with potential application in biomedical or water sensing fields [1]. Bdour et al. present an investigation of the deflection and structural stability of nanohole array-based optofluidic sensors operating in flow-through mode [2]. Hamilton et al. describe a study where 3D hydrodynamic focusing was implemented in 10 µm scale microchannel cross-sections made with a single sacrificial layer, which, implemented in optofluidic sensors, enable higher detection sensitivity and sample specificity [3]. Li et al. report a paper-based photocatalyst immobilization method with enhanced purification efficiency that solves the so-called "coffee ring effect" that occurs on the substrate during solvent evaporation, resulting in the aggregation of the photocatalysts [4]. Ghifari et al. describe a proof of concept of an optofluidic method based on dye-doped ZnO microcapsules for high throughput fluorescence-based thermometry, which enables the measure of temperature inside optofluidic microsystems at the millisecond time scale [5]. Deng et al. report on two kinds of electro-fluidic dyes based on anthraquinone and azo pyrazolone, including their synthesis, structure characterization, and application properties [6]. Yi et al. present a study related to the aperture ratio improvement by optimizing the voltage driving waveform for electrowetting displays [7]. Winskas et al. demonstrate a new bi-metallic substrate that allows micro-scale optofluidic manipulation controlled by an external laser power [8]. Liang et al. present a variety of differently structured optoelectrokinetic chips, discussing how they are fabricated and the ways in which they work. They also provide a summary of the current challenges of optoelectrokinetics and their future prospects [9]. Finally, Wang et al. describe the applications of optofluidic platforms on autonomous and in situ ocean environmental monitoring, with an emphasis on their principles, sensing properties, advantages, and disadvantages [10].

Funding: We thank the AEI-MICINN and EU-FEDER (PID2019-110430GB-C21 and PID2019-107335RA-I00) for financial support.

Acknowledgments: We would like to take this opportunity to thank all the authors for submitting their papers to this Special Issue, all the reviewers for dedicating their time and helping to improve the quality of the submitted papers, and our assistant editor Mandy Zhang for her kind invitation and support for the production of this Special Issue.

Conflicts of Interest: The authors declare no conflict of interest.

References

1. Winskas, J.; Wang, H.; Zhdanov, A.; Cheemalapati, S.; Deonarine, A.; Westerheide, S.; Pyayt, A. Different Regimes of Opto-fluidics for Biological Manipulation. *Micromachines* **2019**, *10*, 802. [CrossRef]
2. Yi, Z.; Feng, W.; Wang, L.; Liu, L.; Lin, Y.; He, W.; Shui, L.; Zhang, C.; Zhang, Z.; Zhou, G. Aperture Ratio Improvement by Optimizing the Voltage Slope and Reverse Pulse in the Driving Waveform for Electrowetting Displays. *Micromachines* **2019**, *10*, 862. [CrossRef] [PubMed]
3. Wang, F.; Zhu, J.; Chen, L.; Zuo, Y.; Hu, X.; Yang, Y. Autonomous and In Situ Ocean Environmental Monitoring on Optofluidic Platform. *Micromachines* **2020**, *11*, 69. [CrossRef] [PubMed]
4. Liang, W.; Liu, L.; Wang, J.; Yang, X.; Wang, Y.; Li, W.; Yang, W. A Review on Optoelectrokinetics-Based Manipulation and Fabrication of Micro/Nanomaterials. *Micromachines* **2020**, *11*, 78. [CrossRef] [PubMed]
5. Deng, Y.; Li, S.; Ye, D.; Jiang, H.; Tang, B.; Zhou, G. Synthesis and a Photo-Stability Study of Organic Dyes for Electro-Fluidic Display. *Micromachines* **2020**, *11*, 81. [CrossRef]
6. Ghifari, N.; Rassouk, S.; Hayat, Z.; Taleb, A.; Chahboun, A.; El Abed, A. Dye-Doped ZnO Microcapsules for High Throughput and Sensitive Optofluidic Micro-Thermometry. *Micromachines* **2020**, *11*, 100. [CrossRef] [PubMed]
7. Li, Q.; Lin, H.; Huang, X.; Lyu, M.; Zhang, H.; Zhang, X.; Wang, R. Paper-based Photocatalysts Immobilization without Coffee Ring Effect for Photocatalytic Water Purification. *Micromachines* **2020**, *11*, 244. [CrossRef] [PubMed]
8. Hamilton, E.; Ganjalizadeh, V.; Wright, J.; Schmidt, H.; Hawkins, A. 3D Hydrodynamic Focusing in Microscale Optofluidic Channels Formed with a Single Sacrificial Layer. *Micromachines* **2020**, *11*, 349. [CrossRef] [PubMed]
9. Bdour, Y.; Gomez-Cruz, J.; Escobedo, C. Structural Stability of Optofluidic Nanostructures in Flow-Through Operation. *Micromachines* **2020**, *11*, 373. [CrossRef] [PubMed]
10. Zuo, Y.; Chen, L.; Hu, X.; Wang, F.; Yang, Y. Silver Nanoprism Enhanced Colorimetry for Precise Detection of Dissolved Oxygen. *Micromachines* **2020**, *11*, 383. [CrossRef] [PubMed]

Review

A Review on Optoelectrokinetics-Based Manipulation and Fabrication of Micro/Nanomaterials

Wenfeng Liang [1], Lianqing Liu [2,3,*], Junhai Wang [1], Xieliu Yang [1], Yuechao Wang [2,3], Wen Jung Li [3,4,*] and Wenguang Yang [5]

[1] School of Mechanical Engineering, Shenyang Jianzhu University, Shenyang 110168, China; liangwf@sjzu.edu.cn (W.L.); jhwang@sjzu.edu.cn (J.W.); yang.xieliu@sjzu.edu.cn (X.Y.)
[2] State Key Laboratory of Robotics, Shenyang Institute of Automation, Chinese Academy of Sciences, Shenyang 110016, China; ycwang@sia.cn
[3] CAS-CityU Joint Laboratory on Robotics, City University of Hong Kong, Kowloon Tong, Hong Kong 999077, China
[4] Department of Mechanical Engineering, City University of Hong Kong, Kowloon Tong, Hong Kong 999077, China
[5] School of Electromechanical and Automotive Engineering, Yantai University, Yantai 264005, China; yangwenguang@ytu.edu.cn
* Correspondence: lqliu@sia.cn (L.L.); wenjli@cityu.edu.hk (W.J.L.); Tel.: +86-24-2397-0181 (L.L.); +852-3442-9266 (W.J.L.)

Received: 20 November 2019; Accepted: 8 January 2020; Published: 10 January 2020

Abstract: Optoelectrokinetics (OEK), a fusion of optics, electrokinetics, and microfluidics, has been demonstrated to offer a series of extraordinary advantages in the manipulation and fabrication of micro/nanomaterials, such as requiring no mask, programmability, flexibility, and rapidness. In this paper, we summarize a variety of differently structured OEK chips, followed by a discussion on how they are fabricated and the ways in which they work. We also review how three differently sized polystyrene beads can be separated simultaneously, how a variety of nanoparticles can be assembled, and how micro/nanomaterials can be fabricated into functional devices. Another focus of our paper is on mask-free fabrication and assembly of hydrogel-based micro/nanostructures and its possible applications in biological fields. We provide a summary of the current challenges facing the OEK technique and its future prospects at the end of this paper.

Keywords: optoelectrokinetics; optically-induced dielectrophoresis; micro/nanomaterials; separation; fabrication

---

1. Introduction

Accurate manipulation and fabrication of micro/nanomaterials in liquid is fundamental for a range of applications such as micro/nanoelectronics [1,2], biosensors [3,4], biomedicine [5,6], biosensing [7,8] and energy harvesting [9,10]. Various attempts have been made to make that happen. For example, silver nanoparticles were integrated into the luminol system to enable more efficient electrochemiluminescence and thereby allow for ultrasensitive detection of cardiac troponin [11]. Combining graphene and traditional integrated circuits, a high-mobility, high-resolution, and broadband image digital sensor was developed to capture ultraviolet, visible, and infrared light [12]. Gold nanoparticles were assembled into micro/nanowires to fabricate a flexible pressure sensor that offers a detection limit up to 25 Pa [13]. This sensor is also sensitive to pulses in different regions of the human body, offering an approach to facilitating the development of wearable devices.

To address real-world needs for manipulation and fabrication of micro/nanomaterials, a number of micro-/nano-scaled methods have been presented. Typical examples include microfluidic [14,15], acoustic [16,17], electrokinetics [18,19], magnetic [20,21], optical [22,23], thermal [24,25], and atomic force microscope [26,27] approaches. An emerging topic on the manipulation and fabrication of micro/nanoparticles is about how to develop a novel mechanism that can complement what a single technique can offer. The long-term aim is to create valuable techniques that can help extend the applicability of micro/nanomaterials. Combining acoustic and magnetic fields, an approach that can aggregate nanoparticles in the presence of a magnetic field and move them towards the wall of a microchannel under the action of an external acoustic wave was proposed [28]. This approach bears significant potential for application in target drug delivery and biomedical surgery. There is also a technique called the acoustofluidic tweezers, which integrates acoustic waves into a microfluidic system and offers a label-free and high-throughput way to isolate 110-nm particles from a mixture of micro/nanoparticles with a purity as high as 99% [29]. By incorporating white light source from light-emitting diode into a microfluidic system, the absorbance of micro/nanoparticles at different diameters as well as the bit error rate can be determined, which allows easy and rapid measurement of micro/nanoparticle concentration [30]. In addition, a hybrid isomotive dielectrophoresis (DEP)-microfluidic technique was proposed for label-free separation of same-/differently-sized micro/nanoparticles based on their crossover-frequency features [31]. This technique also enables reliable and repeatable separation and dynamically adjusting the purity and yield of separated beads.

A new hybrid and novel manipulation technique called optically-induced dielectrophoresis (ODEP) or optoelectrokinetics (OEK) was proposed for programmable, contact-free, flexible, automatic, dynamic, and rapid manipulation of micro/nano-entities [32]. OEK has attracted much attention in micro/nanomanipulation fields since it was invented in 2005. This technique combines the merits of optical, electrokinetics, and microfluidic schemes, offering a more versatile approach to micro-/nano-scaled manipulation and fabrication over other competing lab-on-a-chip techniques. Optically-projected images are custom-designed on a computer using graphic animation software. These images are then transmitted by a commercial projector and focused onto the surface of the OEK chip, thereby triggering the photosensitive material and changing the distribution of the external AC bias potential. Most of the AC bias potential will shift to the liquid layer, which generates a non-uniform electric field around the illuminated area. Meanwhile, various electrokinetics forces are produced. These forces are further exerted onto the micro/nanoparticles, driving, directing, and delivering them towards the intended destination programmatically and digitally. This is how optical-electrokinetics-microfluidics integration works. Featuring the use of optically-projected images as virtual electrodes to directly manipulate micro/nanoparticles and without referring to metal-based electrodes, OEK has been widely used in the manipulation, separation, assembly, and fabrication of micro/nano-entities as well as extraction of their intrinsic properties [33–38], which has shown a highly promising new strategy in the development of micro/nanomanipulation and fabrication communities. More recently, Berkeley Lights Inc., the inventor of OEK, has commercialized this technology in bio-related fields, offering a unique approach to take the development of biomedical and bioengineering sciences to a new level. Furthermore, it also shows the high potential for the use of this technology in material manipulation and fabrication to move from lab-oriented research to real-world applications.

Unlike previous review studies that only discuss some aspects of OEK-based microfluidics [39–41], we present a comprehensive review of OEK-based manipulation and fabrication of micro/nanoparticles. First, we summarize differently structured OEK chips and discuss their respective working principles. Then, we describe OEK-based manipulation and fabrication of micro-/nano-scaled materials without using any masks. Next, we review the use of OEK chips for fabricating various hydrogel-based structures and functional micro/nanodevices. Finally, we explain current and future challenges facing the OEK technique and possible innovations on it.

2. Optoelectrokinetics (OEK) Chips and Their Working Principles

The ODEP or OEK chip was firstly invented in 2005 [32]. The chip typically consists of four layers [32]: a top glass layer coated with a transparent and conductive indium tin oxide (ITO) film, which connects with one end of the AC bias potential; a bottom ITO glass layer serving as the bottom electrode that connects with the other end of the AC bias potential; a thin film (50 nm) n+ hydrogenated amorphous silicon (a-Si:H) layer, deposited onto the bottom ITO glass after generation of a 1 μm undoped a-Si:H layer and a 20 nm silicon nitride layer; a liquid layer that contains the manipulated micro/nanoparticles to function as a "gap" for the OEK chip. The doped a-Si:H layer can decrease the contact resistance between the liquid layer and the undoped a-Si:H layer, and the silicon nitride layer serves the purpose of passivation. It is worth noting that the undoped a-Si:H layer is the photosensitive layer from which optically-induced virtual electrodes are generated. In general, the ITO layer is sputtered onto the glass layer; the a-Si:H layer is deposited onto the ITO layer by plasma-enhanced chemical vapor deposition. Figure 1 is an illustration of its mechanism proposed in our group given in [42]. In our study, only a layer of a-Si:H was deposited, which was also classified into an a-Si:H based OEK chip.

When the OEK chip is illuminated by the optically-projected patterns shown on a digital micromirror display (DMD), the external AC voltage shifts to the liquid layer with suspended micro/nanoparticles. Then, an optically-induced non-uniform electric field is produced around the illuminated area. Under the action of this electric field, the micro/nanoparticles are polarized, producing an ODEP force. This force is, in turn, exerted onto the micro/nanoparticles, which is defined as [43]

$$\left\langle \vec{F}_{ODEP} \right\rangle = 2\pi R^3 \varepsilon_m \mathrm{Re}[K(\omega)] \nabla \left| \vec{E}_{rms} \right|^2, \tag{1}$$

where R is the radius of the micro/nanoparticles, ε is the permittivity of the liquid, E is the magnitude of the optically-induced non-uniform electric field, rms means the root-mean-square value, the subscript m is the liquid, ω is the angular frequency of the AC bias potential across the liquid medium, and $K(\omega)$ is the Clausius–Mossotti (CM) factor, which is further expressed as

$$K(\omega) = \frac{\varepsilon_p^* - \varepsilon_m^*}{\varepsilon_p^* + 2\varepsilon_m^*}, \tag{2}$$

where $\varepsilon^* = \varepsilon - j\sigma/\omega$, with σ representing the conductivity. The subscript p means the particle. The direction of the ODEP force, both positive and negative, is fully directed by the real part of the CM factor. Specifically, if the value of the real part of the CM factor is higher than zero, a positive ODEP force will be generated and then the micro/nanoparticles will be attracted and moved to the illuminated area. Otherwise, the micro/nanoparticles will be pushed away from the illuminated area. For micro/nanoparticles that vary significantly in their size, a common method to separate them is by making the ODEP magnitudes have three orders of difference.

Other OEK forces, including AC electroosmosis (ACEO) and AC electrothermal, were investigated using the finite element analysis (FEA) method [44]. Typically, those two kinds of electrokinetics forces arise from the interaction between the optically-induced non-uniform electric field and liquid solution at a given AC frequency. As an example, under the action of an ACEO force, the fluid flow can rapidly push a large number of micro/nanoparticles towards the illuminated area and assemble them into desired patterns or devices [45,46].

Recently, various types of photosensitive materials have been explored to manipulate and fabricate micro/nanomaterials in the same way as an a-Si:H based OEK chip works. A P3HT/PCBM bulk-heterojunction polymer was demonstrated as an alternative to a-Si:H for micro/nanomanipulation [47]. Figure 2 shows the structure of an OEK chip made up of this material. P3HT/PCBM serves as a photosensitive material to respond to optically-projected images and produce a non-uniform electric field, thereby manipulating and isolating differently sized polystyrene beads [48].

The P3HT/PCBM polymer for making this type of chip is created by spinning-coating at a relatively low temperature. Compared to a-Si:H based OEK chips, OEK chips based on this material are much easier to fabricate, except that the fabrication process of the latter requires preventing the polymer film from being collapsed by either water or oxygen. OEK chips based on this material can bend up and down freely to produce concave or convex curvatures, which leads to significantly higher efficiency in separating and concentrating differently sized polystyrene beads [49].

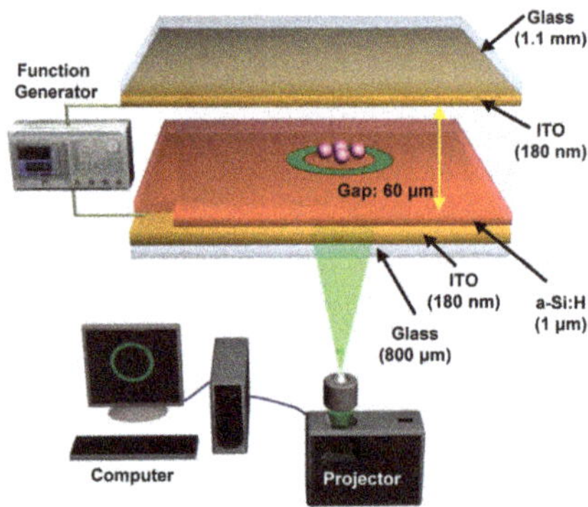

Figure 1. Schematic illustration of optoelectrokinetics (OEK) chip (reproduced from Ref. [42]).

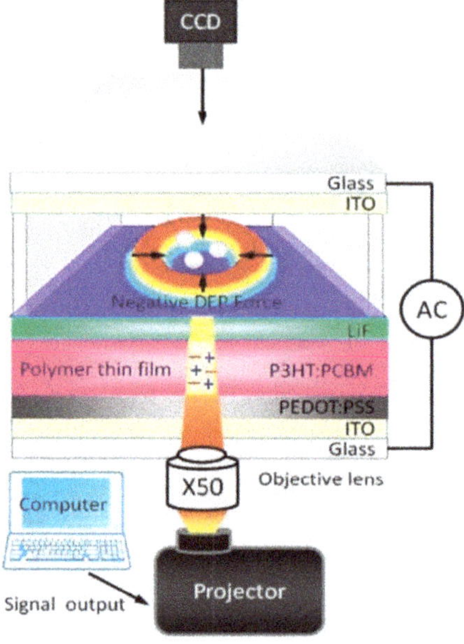

Figure 2. Structure of P3HT/PCBM polymer-based OEK chip (reproduced from Ref. [47]).

Another kind of OEK chip was devised based on T_iOP_c, an organic photosensitive material. As shown in Figure 3 [50], the chip has the same structure as the one indicated in Figure 1. This chip can be easily fabricated only by spinning-coating and baking techniques. This chip is fabricated in a much easier way than an a-Si:H based chip. It has been demonstrated that this chip can perform real-time manipulation of picobubbles [51] and droplets [52] as well as cell patterning [53]. However, problems remain with the long-term stability of this OEK chip.

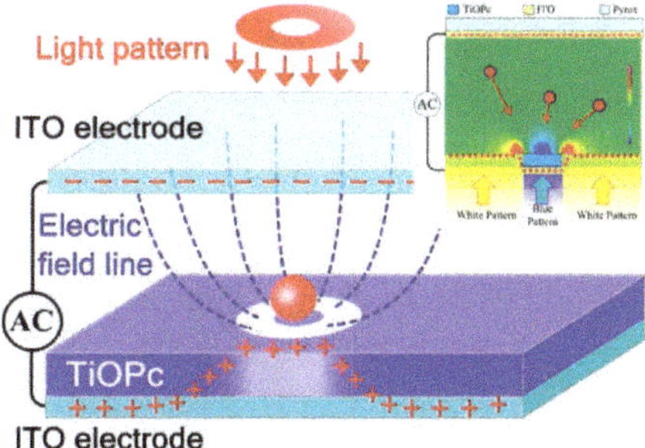

Figure 3. Structure of T$_i$OP$_c$-based OEK chip and simulation of the electric field (reproduced from Ref. [50]).

3. OEK-Based Manipulation of Micro/Nanomaterials

3.1. Separation and Assembly of Micro-Scaled Particles

The OEK chip has been commonly used to perform manipulations such as separation, concentration and assembly on micro/nanoparticles, with the aid of an ODEP and/or ACEO mechanism. The OEK chip permits the assembly of 2D colloidal microparticles by using the electrohydrodynamic flows [54–56] and the manipulation of droplets [57–59]. Differently sized polystyrene beads serve various functions in material science and biomedical research. Hence, finding a rapid and automatic method to manipulate and separate them is critical in investigating the performance of targets of interest.

The conductivity of polystyrene beads is size-dependent and expressed as $\sigma = 2K_S/R$, where K_S is the surface conductivity of polystyrene beads [42]. The relationship between the crossover frequency (Re $[K(\omega)] = 0$) of the ODEP force and the size of the polystyrene beads can be further derived as [42]

$$f_{crossover} = \frac{1}{2\pi} \sqrt{\frac{(\sigma_m R - 2K_S)(2K_S + 2\sigma_m R)}{R^2(\varepsilon_p - \varepsilon_m)(\varepsilon_p + 2\varepsilon_m)}}.$$

(3)

The size-dependent crossover frequency as a function of the liquid conductivity, i.e., Equation (3), is illustrated in Figure 4 [42]. Then, separation of 1 µm and 10 µm polystyrene beads was successfully demonstrated by using ODEP forces in different directions, i.e., a positive ODEP force exerted onto the 1 µm polystyrene beads and a negative one onto the 10 µm ones, as shown in Figure 5 [42].

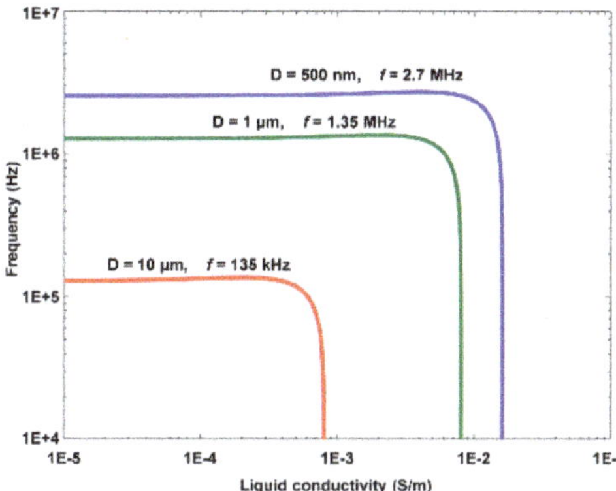

Figure 4. Crossover frequency vs. liquid conductivity of three differently sized polystyrene beads (reproduced from Ref. [42]).

Moreover, polystyrene beads with three different diameters, i.e., 500 nm, 1 µm, and 10 µm, were separated simultaneously. Figure 6 shows the detailed experimental process [42]. A negative ODEP force would be exerted onto the 10 µm diameter polystyrene beads, while a positive ODEP force would be exerted onto both the 500 nm and 1 µm ones. However, the 1 µm polystyrene beads would experience a much higher magnitude of ODEP force than the 500 nm ones because the magnitude of the ODEP force is proportional to the third power of the particle radius.

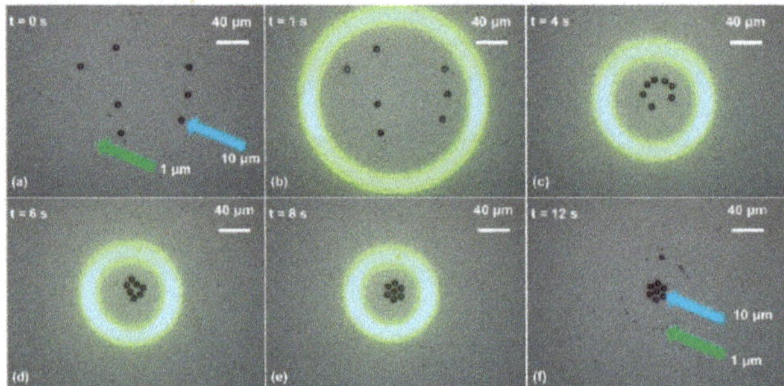

Figure 5. Separation of 10 µm and 1 µm polystyrene beads. (**a**) Initially, polystyrene beads were suspended in the liquid solution; (**b**) the OEK chip was illuminated by the optical ring pattern, with the AC bias potential switched on simultaneously; (**c**) 10 µm polystyrene beads were pushed towards the central area of the ring under the action of a negative ODEP force, while the 1 µm ones were attracted into the ring under the action of a positive ODEP force as the ring size decreased; (**d,e**) are captured images showing the positions of the polystyrene beads with two different diameters as the ring size decreased dynamically; (**f**) shows the final positions of the 10 µm and 1 µm polystyrene beads (reproduced from Ref. [42]).

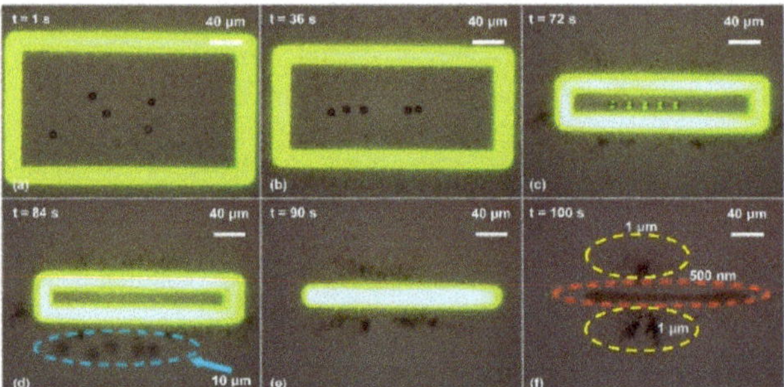

Figure 6. Simultaneous separation of polystyrene beads with diameters of 500 nm, 1 µm, and 10 µm. (**a**) When the OEK chip was illuminated by the optical rectangle pattern under the action of an external AC bias potential, both the 500 nm and 1 µm polystyrene beads were attracted to the illuminated area. By contrast, the 10 µm ones were pushed away from the optical ring; (**b**) as the optical rectangle size decreased, the 10 µm polystyrene beads moved towards the central area of the optical rectangle and those with the other two diameters still resided within the optical ring; (**c**) when the gap formed by the long sides of the rectangle was around 10 µm, the 10 µm polystyrene beads were aligned; (**d**) the 10 µm polystyrene beads were separated as the rectangle size further decreased; (**e**) the rectangle was decreased to a line pattern; (**f**) the 500 nm and 1 µm polystyrene beads were finally located at different positions of the OEK chip (reproduced from Ref. [42]).

Manipulation and assembly of metallic microspheres into patterns were proposed [60]. It was observed that conductive silver-coated poly(methyl methacrylate) (PMMA) microspheres (50 µm diameter) could move in an OEK chip at a maximum velocity of 3200 µm/s, much quicker than non-conductive microparticles such as polystyrene beads [61]. As the motorized XY stage moves at an increasing velocity, the microspheres reach their maximum velocity when they cannot follow the movement of the stage. Simulation on the electric field distribution and the ODEP force attributes the strong ODEP force to the local interaction between the optically-induced electric field and the silver shells surrounding the microspheres. The microspheres were then experimentally assembled to validate their high-accuracy positioning capabilities. Figure 7 shows how the microspheres were assembled with different spaces, i.e., from 1.39 µm to 20.7 µm, relative to a stable bubble [60]. The assembly of the "O," "E," "T" pattern was realized, demonstrating the ability of the OEK chip to precisely and parallelly pattern metallic microspheres into arbitrary shapes. In addition, the manipulation and assembly of hybrid metal-polymer microparticles were achieved [62].

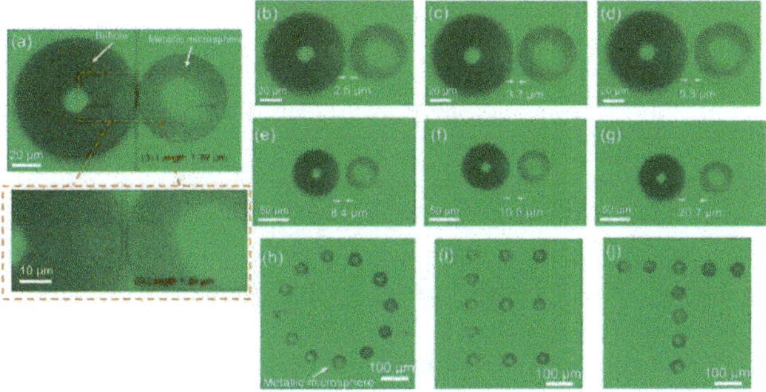

Figure 7. Captured images of manipulated metallic microparticles with different spaces to a bubble: (**a**) 1.39 µm, (**b**) 2.5 µm, (**c**) 3.3 µm, (**d**) 5.3 µm, (**e**) 8.4 µm, (**f**) 10.6 µm, and (**g**) 20.7 µm; parallel assembly of metallic microspheres into images of (**h**) "O," (**i**) "E," (**j**) "T" (reproduced from Ref. [60]).

It has been further demonstrated that OEK can serve as a versatile and programmable microrobot to perform typical micromanipulations such as "load," "transport," and "deliver" [63]. Firstly, OEK was used to manipulate custom-designed microstructures. Then, the OEK based microrobot could manipulate secondary microparticles in a parallel, multistep, contact-free and programmable manner. Figure 8 is a series of images showing the dynamic process of the OEK-based microrobot across large distances. This partially enclosed microrobot could load one 15 µm diameter polystyrene bead of interest with the aid of a negative ODEP force (Figure 8A–C), followed by a translational movement of 300 µm/s (Figure 8D). In addition, the OEK-based microrobot delivered this bead to the target location (Figure 8E,F). Figure 8G schematically presents the three-step process, which exhibited a higher moving velocity than when OEK was used alone. This OEK-based microrobot was also validated to be suitable for isolating single cells for clonal expansion and other biomedical applications.

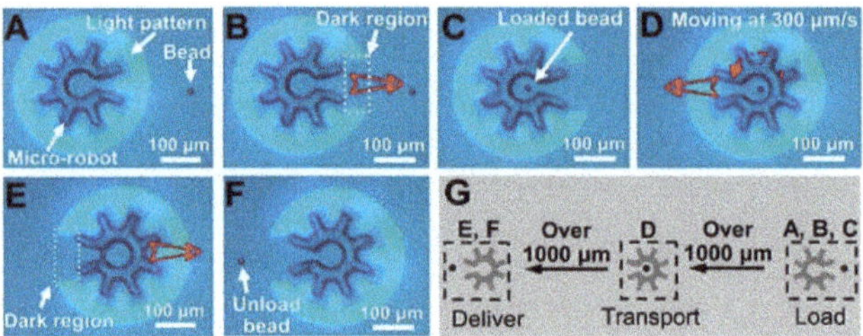

Figure 8. A series of OEK-based robotic micromanipulations. (**A**) A fully enclosed microrobot; (**B**) the load mode of a partially enclosed microrobot; (**C**) the partially enclosed microrobot loaded one bead of interest; (**D**) the fully enclosed microrobot transported the bead; (**E**) the partially enclosed microrobot delivered the bead; (**F**) the bead was unloaded; (**G**) schematic illustration of the "load," "transport," and "deliver" micromanipulations (reproduced from Ref. [63]).

3.2. Manipulation of Nano-Scaled Particles

OEK has also been used to dynamically manipulate nano-scaled entities, including the separation of nanowires [64–67], the patterning of two-dimensional nanomaterials [68], and manipulation of nanoparticles [69–74].

OEK allows the use of ODEP forces to effectively separate individual nanowires with different conductance levels, which has facilitated the development of nanodevices. Dynamic separation of semiconducting and metallic nanowires based on their difference in translational velocity was reported [64]. When the external voltage was higher than the "separation voltage," i.e., the threshold one for moving the silicon nanowire, the silver nanowire could be separated from the silicon nanowire. Due to the high polarization of metallic nanowires in a non-uniform electric field, the silver nanowire moved at a much higher velocity than the silicon nanowire. When the OEK chip was dynamically illuminated by an optically-projected laser line at a scanning velocity greater than 2 μm/s and an AC bias potential of 8 V_{pp}, both of these two types of nanowires were trapped by the laser line initially. The silicon nanowire, however, could not follow the movement of the laser line and became "uncontrollable." Hence, these two types of nanowires were successfully separated. In addition, it was experimentally validated that real-time and large-scale assembly of silver nanowires is possible using an array of image-defined traps, suggesting the potential for massively parallel assembly at the nanoscale.

Furthermore, the manipulation and assembly of nanoparticles have been studied both theoretically and experimentally. First, a rapid and automatic assembly of 100 nm diameter gold nanoparticle (AuNP)-based microstructures was experimentally investigated [72]. Figure 9 indicates the experimental process in which AuNP-based microstructures were rapidly assembled. It was observed that each geometry of the four AuNP-based microstructures could reflect the optical pattern. However, using the triangle pattern as the virtual electrodes attracted most of the AuNPs into the locations of angular bisectors and formed a circle structure with three lines pointing toward the center of the triangle. The square pattern pushed the AuNPs to the diagonals of the pattern and most of them were moved into the central area.

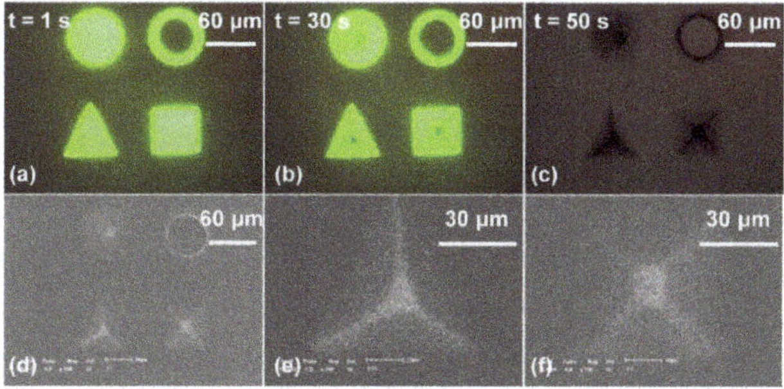

Figure 9. Rapid assembly of various microstructures using gold nanoparticles (AuNPs) with a diameter of 100 nm. (**a**) Four different optical patterns served as virtual electrodes, with an external AC bias potential switched on simultaneously; (**b**) after 30 s, the AuNPs were attracted into the areas illuminated by the four patterns; (**c**) AuNP-based microstructures formed when the optical patterns were moved and AC bias potential was switched off; (**d**) SEM images of (**c**); (**e**) SEM image of an AuNP-based microstructure in triangle pattern; (**f**) SEM image of an AuNP-based microstructure in square pattern (reproduced from Ref. [72]).

Fabrication of various nanomaterial-based microelectrodes in an OEK chip was achieved using two different methods [73]. A composite solution consisting of conductive polyaniline (PANI) nanoparticles and multi-walled carbon nanotubes (MWCNTs) was used to examine the possibility of creating microelectrodes by OEK forces. Microelectrodes with various geometrical sizes could be fabricated within 1–3 min. Compared to the resistance change of the MWCNTs bundles, that of the microelectrodes could be neglected; the ethanol concentration was successfully reflected by the resistance change of the sensitive elements.

Rapid assembly of carbon nanoparticles (CNPs) with a diameter of 50 nm into electrical elements was realized [74]. A series of experiments were conducted to rapidly assemble CNPs into various electrical elements within 45 s. The results demonstrated that the microstructures had resistance properties. Specifically, their resistance value could be controlled by the width and length of the microstructures: inversely increasing with the width (Figure 10a–c) and linearly increasing with the length (Figure 10d–f).

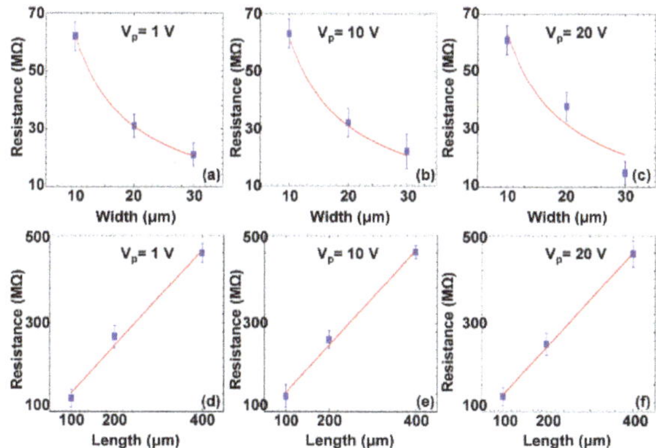

Figure 10. The curve-fitting function of the assembled carbon nanoparticle (CNP)-based microstructures. Resistance with respect to different widths of three microstructures at three different measurement voltages (**a**–**c**); Resistance with respect to different lengths of three microstructures at three different measurement voltages (**d**–**f**) (reproduced from Ref. [74]).

4. Mask-Free Fabrication of Electrodes and Devices

An optically-induced electrochemical reaction and deposition scheme was presented to enable dynamic, rapid and mask-free fabrication of microelectrodes [75–79]. When the OEK chip was illuminated by optically-projected images, an electrical field would be produced in the illuminated area due to the creation of electron-hole pairs. Then, only the metal ions of liquid solution in the illuminated area were reduced by trapping electron from a-Si:H when an external AC bias potential was switched on. Furthermore, the reduced metal atoms would be attached to the illuminated a-Si:H surface and formed into metallic microstructures with the same shape as the incident light.

Compared to the OEK-force-based method, this scheme could dynamically fabricate microelectrodes in 10 secs with a liquid conductivity as high as 2×10^7 S/m. For instance, silver ion-based microelectrodes were reported, as shown in Figure 11 [75]. These microelectrodes exhibited more even distribution and lower roughness. Their heights could be adjusted by the linearly increasing deposition time and the solution concentration. In addition, the entire experiment was conducted at room temperature and atmospheric pressure without using conventional photolithographic techniques and metal nanoparticles and/or inks. In addition, integrated CuO/ZnO-based/single-walled-nanotube [76] nanowire-based field-effect transistors were presented, further validating that this scheme could facilitate the manufacturing of integrated nanodevices and offer an alternative to micro/nanosensor fabrication.

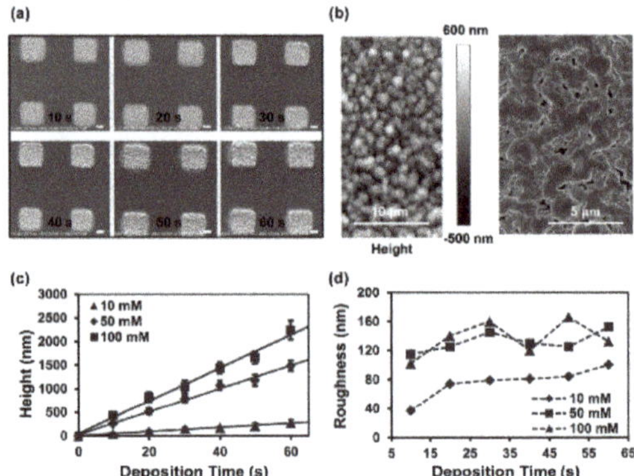

Figure 11. (**a**) Fabricated microelectrodes with respect to deposition time; scale bar: 10 μm. (**b**) Surface roughness of microelectrode by atomic force microscope (AFM) and SEM. (**c**) Thickness and (**d**) surface roughness of microelectrodes with respect to deposition time and solution concentration (reproduced from Ref. [75]).

Silver-based nanostructures with various geometrical topographies were synthesized in an OEK chip, as shown in Figure 12 [77]. The incident light excited the generation of electron-hole pairs in the a-Si:H layer and then the electrons migrated from the valence band to the conduction band. With the aid of an existing electric double layer, an electrochemical reaction occurred between the electrons and the suspended silver solution in the liquid chamber under given parameters in the solution; meanwhile, crystallization occurred with the silver-based nanostructures. In this case, silver polyhedral nanoparticles, nanoplates in triangle and hexagon patterns, as well as nanobelts were fabricated by projecting various optical patterns. Experimental parameters that affected the fabrication process, including time duration, AC frequency and voltage, were also investigated and optimized. In sum, this OEK-based method opens up a new path to mask-free and rapid synthesis of nanostructures and nanobelts.

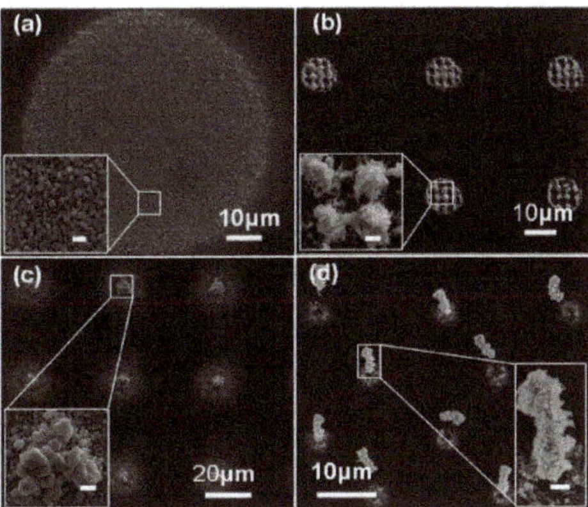

Figure 12. Synthesis of various silver nanostructures by optically-projected patterns in OEK chip. (**a**) Silver crystal nanoparticles. (**b**) Stacked silver hexagonal nanoplates. (**c**) Crystallized silver octahedra and hexagonal silver nanoplates. (**d**) An array of silver nanoparticles-based micropillars (reproduced from Ref. [77]).

Using this method, MoS_2 with excellent optical and electronic properties was fabricated into thin-film transistors without relying on any conventional microlithography, such as nanoimprint lithography, laser patterning, or photolithography [78]. The MoS_2 material was first loaded into the OEK chip. Then, the Au and Ag electrodes were rapidly fabricated onto the target MoS_2 film. Accordingly, MoS_2 thin-film transistors with Au and Ag electrodes were obtained, respectively, as shown in Figure 13. These transistors performed the best when they were 30–40 nm thick, exhibiting a low subthreshold swing of 0.75 V/decade and high mobility of 41 $cm^2 \cdot V^{-1} \cdot s^{-1}$, much better than conventional Si-based thin-film transistors.

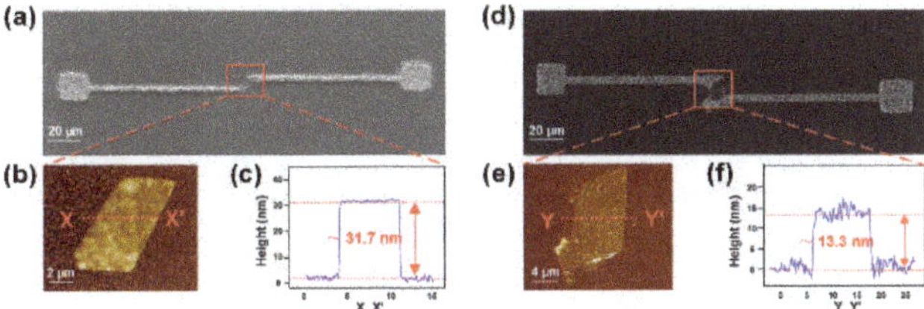

Figure 13. SEM images of MoS_2 thin-film transistors fabricated with Au (**a**) and Ag (**d**) electrodes; (**b**,**e**) are the enlarged view of the rectangle areas in (**a**,**d**), respectively; (**c**,**f**) are the AFM characterization results of the heights of MoS_2 thin-film transistors (reproduced from Ref. [78]).

5. Fabrication and Assembly of Hydrogel-based Micro/Nanostructures

Mask-free and non-UV based polymerization and prototyping of high-aspect-ratio 3D hydrogel microstructures, such as poly (ethylene glycol) (PEG)-diacrylate (PEGDA), was demonstrated in an OEK chip [80–85].

A laser or UV-based method was proposed to obtain PEGDA-based micro/nanostructures [86,87]. The abovementioned OEK-based mask-free method could meet the same purpose without using any UV light sources or lasers [80]. Figure 14 shows the experimental results with controlled sizes, shapes and thicknesses of PEGDA-based structures. Parallel micro/nanostructures were flexibly patterned onto the chip using optical images. Furthermore, by projecting a series of custom-designed and dynamic optical patterns that served as digital masks, 3D microstructures were fabricated rapidly in a layer-by-layer manner, with their thickness varying from tens of nanometers to hundreds of micrometers [81].

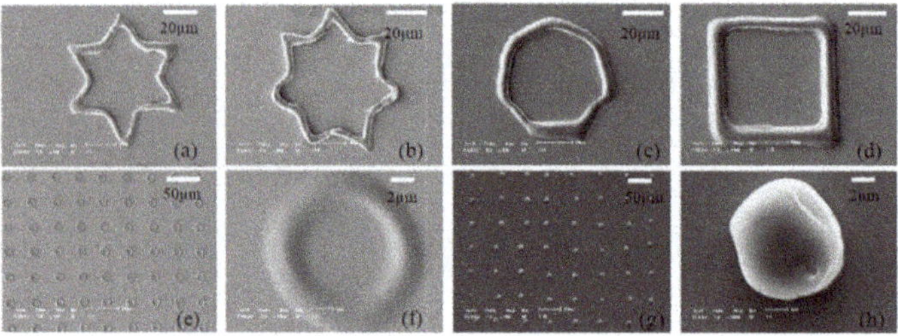

Figure 14. SEM images of poly (ethylene glycol) (PEG)-diacrylate (PEGDA)-based micro/nanostructures fabricated using optically-projected spot patterns with different exposure times. The heights in (a–f) measured by AFM were 233 nm, 455 nm, 687 nm, 950 nm, 1.34 µm, and 1.62 µm, respectively (reproduced from Ref. [80]).

In addition, hollow and circular tubes were fabricated with a length, diameter, wall thickness, and high-aspect-ratio tuned by the exposure time, as shown in Figure 15 [82]. These tubes could manipulate and trap polystyrene beads when they grew longer as the exposure time increased. Then, the PEGDA-based tube continued to elongate and the beads were trapped and moved away from the initial position. Arrays of hydrogel micropillars were also fabricated. These micropillars were capable of serving as micro-scaled cavity molds for casting polydimethylsiloxane, demonstrating potential of finding applications in microfluidics-related fields. The abovementioned results indicate this technology is well-suited for the rapid fabrication of microfluidic chips.

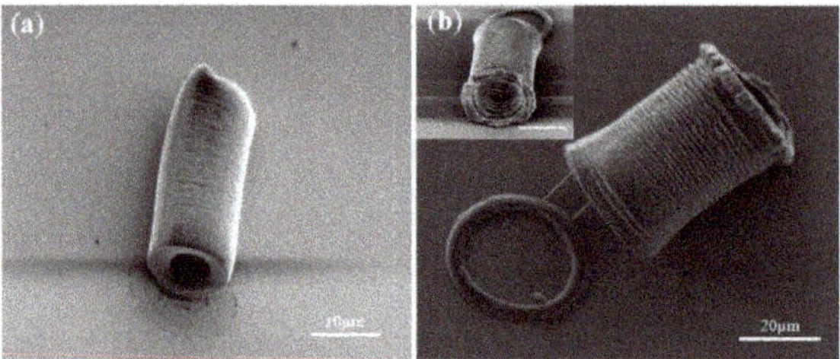

Figure 15. (a) PEGDA-based hydrogel tube with a length of ~20 µm, an outer diameter of ~10 µm, and an inner diameter of ~5 µm. (b) PEGDA-based hydrogel tube with a length of ~35 µm, an outer diameter of ~25 µm, and an inner diameter of ~20 µm (reproduced from Ref. [82]).

Assembly of PEGDA-based microstructures was also realized, followed by an application of bottom-up functional tissue construction and engineering [83]. Arbitrary PEGDA-based microstructures were polymerized in a high-throughput and on-demand manner by a DMD. Then, the fabricated microstructures were transferred into an OEK chip for assembly, which was achieved by using microfluidic flow. Figure 16 illustrates the assembly of PEGDA-based microstructures with different sizes and shapes. Before moving into the OEK chip, the microstructures were either fabricated individually or patterned on a per-array basis. Then, microstructures were aligned into horizontal lines under the action of an ODEP force. Figure 16A–G present the dynamic process of translating microstructures into a single structure for further biomedical and tissue applications. Furthermore, microstructures with the same shape could be assembled into the same layer, as shown in Figure 16H–L.

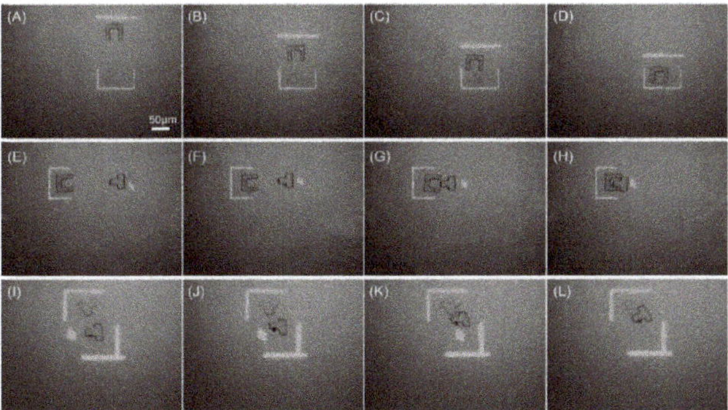

Figure 16. (**A–G**) Captured microscope image showing the dynamic process of assembling microstructures with different sizes and shapes; (**H–L**) the experimental process of assembling two different microstructures (reproduced from Ref. [83]).

An extended study was made on the microfabrication of 3D hydrogel scaffolds [84]. A layer-by-layer solidification mechanism was presented to fabricate hydrogel scaffolds, which involved a polymerization-delamination-polymerization loop. This loop was determined by a competition between the adhesive force and the water-absorbency-induced swelling force. The hydrogel scaffolds' thickness ranged from tens of nanometers to hundreds of nanometers. Thin hydrogel layers were cured at the interface of the a-Si:H and PEDGA solution layers in a layer-by-layer scheme with nano-scaled thickness. Furthermore, those layers were continuously stacked along the normal direction of the OEK chip to finally construct 3D structures. Figure 17 is a microscopic characterization of the dynamic multilayered electro-polymerization of 3D hydrogel structures. Two alternating lines perpendicular to each other were designed to fabricate hydrogel microstructures, as shown in Figure 17a. The optical microscope, SEM, and AFM were employed to illustrate the fabrication results at a Δt_0 of 4 secs, as shown in Figure 17b–d, respectively. The results indicated that these mesh-like hydrogel scaffolds came with controlled pores and gaps. The folded PEGDA hydrogel grids shown in Figure 17e,f further validated the existence of pores and gaps, which could greatly facilitate the spreading, migration, and proliferation of cells as well as the mimicking of cell-cell communications.

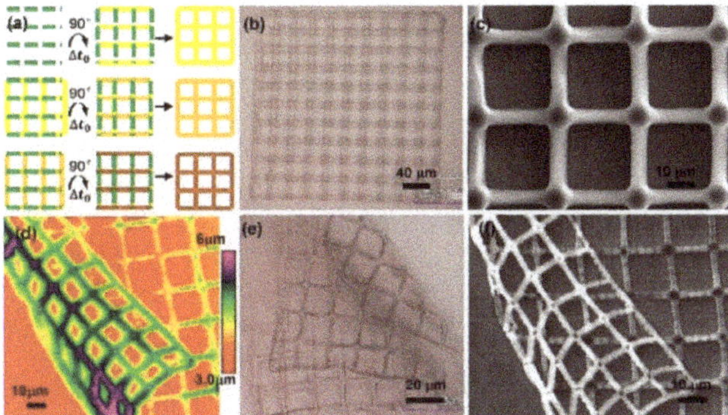

Figure 17. (**a**) Projected patterns alternating at a time interval (Δt_0); (**b–d**) are the optical microscope, SEM and AFM observation results, respectively; (**e,f**) are the optical microscope and SEM images of folded PEGDA hydrogel grids, respectively (reproduced from Ref. [84]).

6. Conclusions and Prospects

As discussed above, the OEK technique has been widely used by the microfluidic community to make a list of research achievements in terms of the manipulation and fabrication of micro/nanomaterials Although some of these achievements are also accomplished by other competing techniques, OEK is superior to the others in that it consumes ultra-low power and offers a programmable, flexible and versatile approach for parallel and multi-scaled micromanipulation. OEK has also been found to be capable of detecting cellular properties and statuses and measuring drug concentration in a label-free and dynamic manner, without relying on any other techniques [88–94]. These advantages suggest that OEK is well-positioned to promote the use of microfluidic tools for micro/nanomanipulation. Nonetheless, challenges remain for OEK to transform from a lab-based technique to one that is widely useful in practical applications. Here is a summary of these challenges.

Firstly, the 3D manipulation mechanism of OEK should be further explored. Early studies on the OEK technique were mostly focused on 2D manipulation schemes, which proved to be more effective than other competing techniques. With the evolvement of OEK chips, however, researchers have realized the necessity of shifting their focus to 3D manipulation. One example of this shift in focus is a two-layer a-Si:H based OEK chip, in which the top ITO glass is replaced with a-Si:H to enable spatial and adsorption-free manipulation of beads using a negative ODEP force [95]. However, this OEK chip cannot manipulate micro/nanoparticles using a positive ODEP force and requires using other OEK forces. Besides, the manipulation is liable to be affected by dynamic fluid flow and electrothermal motion, which is because the manipulated micro/nanoparticles tend to be located in the center of the chip. The poor light transmission can significantly compromise the quality of observation. A lateral-field OEK device was developed to enable parallel single-cell manipulation, which can also assemble nanowires and integrate with other microfluidic components [96]. Nevertheless, the fabrication process is complicated and involves high costs. By dynamically projecting a series of optical patterns, a layer-by-layer approach was built to create 3D microstructures in a mask-free manner [81,84]. However, these incident light patterns are only suitable for fabricating hydrogel microstructures and are insufficient for manipulating nanoparticles and assembling nanowires into functional devices. All the above problems warrant continued efforts to further enhance capabilities of OEKs in 3D manipulation.

Secondly, to facilitate the adoption of OEK-based microfluidics, it is important to develop an integrated chip that is ready for a complete and conventional laboratory-level process. This requires the integration of OEK with conventional microfluidic systems and/or components. A number of attempts have been made on such integration to move OEK further beyond its typical functions such as manipulation, separation, assembly, and fabrication of micro/nano-entities. For instance, to enhance the detection performance of OEK, a pair of optical fibers were embedded into the microchannel to rapidly and accurately count the number of particles and cells [97,98]. A lens-free holographic microscope was incorporated into the OEK chip to allow observing microparticles and cells in a large field of view [99]. This approach offers the ability to rapidly capture the holographic images of microparticles and cells across an ultra-large area of up to hundreds of square millimeters. In addition, on-chip continuous medium exchange and electroporation of cells [100] were achieved by incorporating a conventional microchannel bonded onto a top ITO glass layer. The combination of an OEK chip and surface acoustic wave elements was utilized [101,102] to provide functions such as concentration, guiding, focusing, trapping, and sorting of polystyrene beads and cell lysis. Although the abovementioned works represent tangible advances in the OEK technique, there is still a lack of effective integration of micro/nanochannels or micro/nanostructures into the liquid layer of OEK chips. This severely hinders the use of OEK in fabricating arrays of micro/nanosensors and sophisticated micro/nanodevices for bio-detection and tissue engineering applications. The fabricated micro/nanostructures mentioned in this paper still reside within the OEK chip. They cannot be delivered or directed out of the chip due to the lack of a conventional micropump or microchannel integrated into the liquid layer. Hence, there is much more to do to better integrate OEK chips with microfluidic systems or elements.

Finally, increased efforts are required to identify new possible applications of the OEK technique, which is the ultimate challenge and also the key to address the above two challenges. We believe that a technique can only flourish by bringing real benefits to end-users. Although a variety of OEK-based applications, such as separation, assembly, patterning, fabrication, and synthesis of materials, have been reported, this technique still has a long way to go to become a truly useful and practical tool in these endeavors. For example, there has not been an OEK-based method to fabricate functional micro/nanodevices for industrial applications, such as field-effect transistors and nanosensors. In sum, if upcoming studies can focus more on solving the abovementioned challenges, the OEK technique will soon be able to find wide applications in real-world situations.

Author Contributions: W.L., L.L. and W.J.L. proposed the original idea, and planned the configuration. W.L. wrote the manuscript. J.W., X.Y., Y.W. and W.Y. revised the paper for language and quality. All authors have read and agreed to the published version of the manuscript.

Funding: This work was supported by the National Natural Science Foundation of China (grant numbers 61973224, 51805336, U1613220, U1908215 and 61803323); the Natural Science Foundation of Liaoning Province (grant numbers 2019-KF-01-15 and 2019-ZD-0673); and the Scientific Research Innovation Cultivation Project of Shenyang Jianzhu University (grant number CXPY2017012).

References

1. Slota, M.; Keerthi, A.; Myers, W.K.; Tretyakov, E.; Baumgarten, M.; Ardavan, A.; Sadeghi, H.; Lambert, C.J.; Narita, A.; Müllen, K.; et al. Magnetic edge states and coherent manipulation of graphene nanoribbons. *Nature* **2018**, *557*, 691–695. [CrossRef] [PubMed]

2. He, W.; Qin, C.; Qiao, Z.; Gong, Y.; Zhang, X.; Zhang, G.; Chen, R.; Gao, Y.; Xiao, L.; Jia, S. In situ manipulation of fluorescence resonance energy transfer between quantum dots and monolayer graphene oxide by laser irradiation. *Nanoscale* **2019**, *11*, 1236–1244. [CrossRef] [PubMed]

3. Cui, W.; Mu, L.; Duan, X.; Pang, W.; Reed, M.A. Trapping of sub-100 nm nanoparticles using gigahertz acoustofluidic tweezers for biosensing applications. *Nanoscale* **2019**, *11*, 14625–14634. [CrossRef] [PubMed]

4. Khalil, I.; Yehye, W.A.; Julkapli, N.M.; Rahmati, S.; Ibn Sina, A.A.; Basirun, W.J.; Johan, M.R. Graphene oxide and gold nanoparticle based dual platform with short DNA probe for the PCR free DNA biosensing using surface-enhanced Raman scattering. *Biosens. Bioelectron.* **2019**, *131*, 214–223. [CrossRef]

5. Kyriazi, M.-E.; Giust, D.; El-Sagheer, A.H.; Lackie, P.M.; Muskens, O.L.; Brown, T.; Kanaras, A.G. Multiplexed mRNA Sensing and Combinatorial-Targeted Drug Delivery Using DNA-Gold Nanoparticle Dimers. *ACS Nano* **2018**, *12*, 3333–3340. [CrossRef]

6. Du, J.; Ge, H.; Long, S.; Sun, W.; Fan, J.; Peng, X. Gold nanoparticle-based plasmonic probe for selective recognition of adenosine. *Sens. Actuators B Chem.* **2019**, *296*, 126591. [CrossRef]

7. Sailor, M.J.; Wu, E.C. Photoluminescence-based sensing with porous silicon films, microparticles, and nanoparticles. *Adv. Funct. Mater.* **2009**, *19*, 3195–3208. [CrossRef]

8. Li, N.; Zheng, J.; Li, C.; Wang, X.; Ji, X.; He, Z. An enzyme-free DNA walker that moves on the surface of functionalized magnetic microparticles and its biosensing analysis. *Chem. Commun.* **2017**, *53*, 8486–8488. [CrossRef]

9. Cauda, V.; Stassi, S.; Lamberti, A.; Morello, M.; Pirri, C.F.; Canavese, G. Leveraging ZnO morphologies in piezoelectric composites for mechanical energy harvesting. *Nano Energy* **2015**, *18*, 212–221. [CrossRef]

10. Bhattacharjee, M.; Pasumarthi, V.; Chaudhuri, J.; Singh, A.K.; Nemade, H.B.; Bandyopadhyay, D. Self-spinning nanoparticle laden microdroplets for sensing with energy harvesting. *Nanoscale* **2016**, *8*, 6118–6128. [CrossRef]

11. Wang, S.; Zhao, Y.; Wang, M.; Li, H.; Saqib, M.; Ge, C.; Zhang, X.; Jin, Y. Enhancing luminol electrochemiluminescence by combined use of cobalt-based metal organic frameworks and silver nanoparticles and its application in ultrasensitive detection of cardiac troponin i. *Anal. Chem.* **2019**, *91*, 3048–3054. [CrossRef] [PubMed]

12. Goossens, S.; Navickaitė, G.; Monasterio, C.; Gupta, S.; Piqueras, J.J.; Perez, R.; Burwell, G.; Nikitskiy, I.; Lasanta, T.; Galán, T.; et al. Broadband image sensor array based on graphene–CMOS integration. *Nat. Photon.* **2017**, *11*, 366–371. [CrossRef]

13. Li, S.-X.; Xia, H.; Xu, Y.-S.; Lv, C.; Wang, G.; Dai, Y.-Z.; Sun, H.-B. Gold nanoparticle densely packed micro/nanowire-based pressure sensors for human motion monitoring and physiological signal detection. *Nanoscale* **2019**, *11*, 4925–4932. [CrossRef] [PubMed]

14. Xiang, N.; Wang, J.; Li, Q.; Han, Y.; Huang, D.; Ni, Z. Precise size-based cell separation via the coupling of inertial microfluidics and deterministic lateral displacement. *Anal. Chem.* **2019**, *91*, 10328–10334. [CrossRef] [PubMed]

15. Tian, F.; Feng, Q.; Chen, Q.; Liu, C.; Li, T.; Sun, J. Manipulation of bio-micro/nanoparticles in non-newtonian microflows. *Microfluid. Nanofluidics* **2019**, *23*, 68. [CrossRef]

16. Mao, Z.; Li, P.; Wu, M.; Bachman, H.; Mesyngier, N.; Guo, X.; Liu, S.; Costanzo, F.; Huang, T.J. Enriching NanoparticlesviaAcoustofluidics. *ACS Nano* **2017**, *11*, 603–612. [CrossRef]

17. Zheng, T.; Wang, C.; Xu, C.; Hu, Q.; Wei, S. Patterning microparticles into a two-dimensional pattern using one column standing surface acoustic waves. *Sens. Actuators A Phys.* **2018**, *284*, 168–171. [CrossRef]

18. Park, S.; Yossifon, G. Combining dielectrophoresis and concentration polarization-based preconcentration to enhance bead-based immunoassay sensitivity. *Nanoscale* **2019**, *11*, 9436–9443. [CrossRef]

19. Liu, L.; Chen, K.; Xiang, N.; Ni, Z. Dielectrophoretic manipulation of nanomaterials: A review. *Electrophoresis* **2019**, *40*, 873–889. [CrossRef]

20. Zhang, Y.; DaSilva, M.; Ashall, B.; Doyle, G.; Zerulla, D.; Sands, T.D.; Lee, G.U. Magnetic manipulation and optical imaging of an active plasmonic single-particle Fe–Au nanorod. *Langmuir* **2011**, *27*, 15292–15298. [CrossRef]

21. Ebrahimian, H.; Giesguth, M.; Dietz, K.-J.; Reiss, G.; Herth, S. Magnetic tweezers for manipulation of magnetic particles in single cells. *Appl. Phys. Lett.* **2014**, *104*, 63701. [CrossRef]

22. Nan, F.; Yan, Z. Silver-nanowire-based interferometric optical tweezers for enhanced optical trapping and binding of nanoparticles. *Adv. Funct. Mater.* **2019**, *29*, 1808258. [CrossRef]

23. Gould, O.E.C.; Box, S.J.; Boott, C.E.; Ward, A.D.; Winnik, M.A.; Miles, M.J.; Manners, I. Manipulation and deposition of complex, functional block copolymer nanostructures using optical tweezers. *ACS Nano* **2019**, *13*, 3858–3866. [CrossRef] [PubMed]

24. Manna, R.K.; Shklyaev, O.E.; Kauffman, J.; Tansi, B.; Sen, A.; Balazs, A.C. Light-induced convective segregation of different sized microparticles. *ACS Appl. Mater. Interfaces* **2019**, *11*, 18004–18012. [CrossRef] [PubMed]

25. Tsuji, T.; Matsumoto, Y.; Kugimiya, R.; Doi, K.; Kawano, S. Separation of nano—and microparticle flows using thermophoresis in branched microfluidic channels. *Micromachines* **2019**, *10*, 321. [CrossRef]

26. Li, H.; Han, Y.; Duan, T.; Leifer, K. Size-dependent elasticity of gold nanoparticle measured by atomic force microscope based nanoindentation. *Appl. Phys. Lett.* **2019**, *115*, 053104. [CrossRef]

27. Moreno, M.M.; Ares, P.; Moreno, C.; Zamora, F.; Gomez-Navarro, C.; Herrero, J.G. AFM manipulation of gold nanowires to build electrical circuits. *Nano Lett.* **2019**, *19*, 5459–5468. [CrossRef]

28. Ahmed, D.; Baasch, T.; Blondel, N.; Läubli, N.; Dual, J.; Nelson, B.J. Neutrophil-inspired propulsion in a combined acoustic and magnetic field. *Nat. Commun.* **2017**, *8*, 770. [CrossRef]

29. Wu, M.; Ouyang, Y.; Wang, Z.; Zhang, R.; Huang, P.-H.; Chen, C.; Li, H.; Li, P.; Quinn, D.; Dao, M.; et al. Isolation of exosomes from whole blood by integrating acoustics and microfluidics. *Proc. Natl. Acad. Sci. USA* **2017**, *114*, 10584–10589. [CrossRef]

30. Zhao, W.; Yu, H.; Wen, Y.; Wang, F.; Yang, Y.; Liu, Z.; Liu, L.; Li, W.J. Detection of micro/nano-particle concentration using modulated light-emitting diode white light source. *Sens. Actuators A Phys.* **2019**, *285*, 89–97. [CrossRef]

31. Shkolnikov, V.; Xin, D.; Chen, C. Continuous dielectrophoretic particle separation via isomotive dielectrophoresis with bifurcating stagnation flow. *Electrophoresis* **2019**, *40*, 2988–2995. [CrossRef]

32. Chiou, P.Y.; Ohta, A.T.; Wu, M.C. Massively parallel manipulation of single cells and microparticles using optical images. *Nature* **2005**, *436*, 370–372. [CrossRef]

33. Yang, X.; Niu, X.; Liu, Z.; Zhao, Y.; Zhang, G.; Liang, W.; Li, W.J. Accurate extraction of the self-rotational speed for cells in an electrokinetics force field by an image matching algorithm. *Micromachines* **2017**, *8*, 282. [CrossRef] [PubMed]

34. Liang, W.; Wang, Y.; Zhang, H.; Liu, L. Characterization of the self-rotational motion of stored red blood cells by using optically-induced electrokinetics. *Opt. Lett.* **2016**, *41*, 2763. [CrossRef]

35. Liang, W.; Zhang, K.; Yang, X.; Liu, L.; Yu, H.; Zhang, W. Distinctive translational and self-rotational motion of lymphoma cells in an optically induced non-rotational alternating current electric field. *Biomicrofluidics* **2015**, *9*, 014121. [CrossRef]

36. Liang, W.; Zhao, Y.; Liu, L.; Wang, Y.; Dong, Z.; Li, W.J.; Lee, G.-B.; Xiao, X.; Zhang, W. Rapid and Label-Free Separation of Burkitt's Lymphoma Cells from Red Blood Cells by Optically-Induced Electrokinetics. *PLoS ONE* **2014**, *9*, e90827. [CrossRef] [PubMed]

37. Ke, L.-Y.; Kuo, Z.-K.; Chen, Y.-S.; Yeh, T.-Y.; Dong, M.; Tseng, H.-W.; Liu, C.-H. Cancer immunotherapy μ-environment LabChip: taking advantage of optoelectronic tweezers. *Lab Chip* **2018**, *18*, 106–114. [CrossRef] [PubMed]

38. Kamata, M.; Taguchi, Y.; Nagasaka, Y. Design of an optofluidic diffusion sensor by transient grating using dielectrophoresis. *Opt. Express* **2018**, *26*, 16970–16983. [CrossRef] [PubMed]

39. Wu, M.C. Optoelectronic tweezers. *Nat. Photon.* **2011**, *5*, 322–324. [CrossRef]

40. Hwang, H.; Park, J.-K. Optoelectrofluidic platforms for chemistry and biology. *Lab Chip* **2011**, *11*, 33–47. [CrossRef] [PubMed]

41. Liang, W.; Liu, L.; Zhang, H.; Wang, Y.; Li, W.J. Optoelectrokinetics-based microfluidic platform for bioapplications: A review of recent advances. *Biomicrofluidics* **2019**, *13*, 051502. [CrossRef] [PubMed]

42. Liang, W.; Liu, N.; Dong, Z.; Liu, L.; Mai, J.D.; Lee, G.-B.; Li, W.J. Simultaneous separation and concentration of micro- and nano-particles by optically induced electrokinetics. *Sens. Actuators A Phys.* **2013**, *193*, 103–111. [CrossRef]

43. Castellanos, A.; Ramos, A.; González, A.; Green, N.G.; Morgan, H. Electrohydrodynamics and dielectrophoresis in microsystems: scaling laws. *J. Phys. D Appl. Phys.* **2003**, *36*, 2584–2597. [CrossRef]

44. Valley, J.K.; Jamshidi, A.; Ohta, A.T.; Hsu, H.-Y.; Wu, M.C. Operational regimes and physics present in optoelectronic tweezers. *J. Microelectrom. Syst.* **2008**, *17*, 342–350. [CrossRef] [PubMed]

45. Chiou, P.-Y.; Ohta, A.; Jamshidi, A.; Hsu, H.-Y.; Wu, M. Light-actuated AC electroosmosis for nanoparticle manipulation. *J. Microelectrom. Syst.* **2008**, *17*, 525–531. [CrossRef]

46. Hwang, H.; Park, J.-K. Rapid and selective concentration of microparticles in an optoelectrofluidic platform. *Lab Chip* **2009**, *9*, 199–206. [CrossRef]

47. Wang, W.; Lin, Y.-H.; Guan, R.-S.; Wen, T.-C.; Guo, T.-F.; Lee, G.-B. Bulk-heterojunction polymers in optically-induced dielectrophoretic devices for the manipulation of microparticles. *Opt. Express* **2009**, *17*, 17603–17613. [CrossRef]

48. Wang, W.; Lin, Y.-H.; Wen, T.-C.; Guo, T.-F.; Lee, G.-B. Selective manipulation of microparticles using polymer-based optically induced dielectrophoretic devices. *Appl. Phys. Lett.* **2010**, *96*, 113302. [CrossRef]

49. Lin, S.-J.; Hung, S.-H.; Jeng, J.-Y.; Guo, T.-F.; Lee, G.-B. Manipulation of micro-particles by flexible polymer-based optically-induced dielectrophoretic devices. *Opt. Express* **2012**, *20*, 583–592. [CrossRef]

50. Yang, S.-M.; Yu, T.-M.; Huang, H.-P.; Ku, M.-Y.; Hsu, L.; Liu, C.-H. Dynamic manipulation and patterning of microparticles and cells by using TiOPc-based optoelectronic dielectrophoresis. *Opt. Lett.* **2010**, *35*, 1959–1961. [CrossRef]

51. Yang, S.-M.; Yu, T.-M.; Huang, H.-P.; Ku, M.-Y.; Tseng, S.-Y.; Tsai, C.-L.; Chen, H.-P.; Hsu, L.; Liu, C.-H. Light-driven manipulation of picobubbles on a titanium oxide phthalocyanine-based optoelectronic chip. *Appl. Phys. Lett.* **2011**, *98*, 153512. [CrossRef]

52. Yu, T.-M.; Yang, S.-M.; Fu, C.-Y.; Liu, M.-H.; Hsu, L.; Chang, H.-Y.; Liu, C.-H. Integration of organic opto-electrowetting and poly(ethylene) glycol diacrylate (PEGDA) microfluidics for droplets manipulation. *Sens. Actuators B Chem.* **2013**, *180*, 35–42. [CrossRef]

53. Yang, S.-M.; Tseng, S.-Y.; Chen, H.-P.; Hsu, L.; Liu, C.-H. Cell patterning via diffraction-induced optoelectronic dielectrophoresis force on an organic photoconductive chip. *Lab Chip* **2013**, *13*, 3893. [CrossRef] [PubMed]

54. Hwang, H.; Park, Y.-H.; Park, J.-K. Optoelectrofluidic control of colloidal assembly in an optically induced electric field. *Langmuir* **2009**, *25*, 6010–6014. [CrossRef] [PubMed]

55. Huang, J.-Y.; Lu, Y.-S.; Yeh, J.A. Self-assembled high NA microlens arrays using global dielectricphoretic energy wells. *Opt. Express* **2006**, *14*, 10779. [CrossRef] [PubMed]

56. Lin, W.-Y.; Lin, Y.-H.; Lee, G.-B. Separation of micro-particles utilizing spatial difference of optically induced dielectrophoretic forces. *Microfluid. Nanofluidics* **2010**, *8*, 217–229. [CrossRef]

57. Park, S.-Y.; Kalim, S.; Callahan, C.; Teitell, M.A.; Chiou, E.P.Y. A light-induced dielectrophoretic droplet manipulation platform. *Lab Chip* **2009**, *9*, 3228. [CrossRef]

58. Lee, H.; Hwang, H.; Park, J.-K. Generation and manipulation of droplets in an optoelectrofluidic device integrated with microfluidic channels. *Appl. Phys. Lett.* **2009**, *95*, 164102. [CrossRef]

59. Hung, S.; Lin, Y.; Lee, G. A microfluidic platform for manipulation and separation of oil-in-water emulsion dielectrophoresis. *J Micromechan. Microeng.* **2010**, *20*, 045026. [CrossRef]

60. Zhang, S.; Juvert, J.; Cooper, J.M.; Neale, S.L. Manipulating and assembling metallic beads with optoelectronic tweezers. *Sci. Rep.* **2016**, *6*, 32840. [CrossRef]

61. Liang, W.; Wang, S.; Dong, Z.; Lee, G.-B.; Li, W.J. Optical spectrum and electric field waveform dependent optically-induced dielectrophoretic (ODEP) micro-manipulation. *Micromachines* **2012**, *3*, 492–508. [CrossRef]

62. Han, D.; Hwang, H.; Park, J.-K. Optoelectrofluidic behavior of metal–polymer hybrid colloidal particles. *Appl. Phys. Lett.* **2013**, *102*, 054105. [CrossRef]

63. Zhang, S.; Scott, E.Y.; Singh, J.; Chen, Y.; Zhang, Y.; Elsayed, M.; Chamberlain, M.D.; Shakiba, N.; Adams, K.; Yu, S.; et al. The optoelectronic microrobot: A versatile toolbox for micromanipulation. *Proc. Natl. Acad. Sci. USA* **2019**, *116*, 14823–14828. [CrossRef] [PubMed]

64. Jamshidi, A.; Pauzauskie, P.J.; Schuck, P.J.; Ohta, A.T.; Chiou, P.-Y.; Chou, J.; Yang, P.; Wu, M.C. Dynamic manipulation and separation of individual semiconducting and metallic nanowires. *Nat. Photon.* **2008**, *2*, 86–89. [CrossRef]

65. Pauzauskie, P.J.; Jamshidi, A.; Valley, J.K.; Satcher, J.H.; Wu, M.C. Parallel trapping of multiwalled carbon nanotubes with optoelectronic tweezers. *Appl. Phys. Lett.* **2009**, *95*, 113104. [CrossRef]

66. Lee, M.-W.; Lin, Y.-H.; Lee, G.-B. Manipulation and patterning of carbon nanotubes utilizing optically induced dielectrophoretic forces. *Microfluid. Nanofluidics* **2010**, *8*, 609–617. [CrossRef]

67. Lin, Y.-H.; Ho, K.-S.; Yang, C.-T.; Wang, J.-H.; Lai, C.-S. A highly flexible platform for nanowire sensor assembly using a combination of optically induced and conventional dielectrophoresis. *Opt. Express* **2014**, *22*, 13811–13824. [CrossRef]

68. Lim, M.B.; Felsted, R.G.; Zhou, X.; Smith, B.E.; Pauzauskie, P.J. Patterning of graphene oxide with optoelectronic tweezers. *Appl. Phys. Lett.* **2018**, *113*, 031106. [CrossRef]

69. Jamshidi, A.; Neale, S.L.; Yu, K.; Pauzauskie, P.J.; Schuck, P.J.; Valley, J.K.; Hsu, H.-Y.; Ohta, A.T.; Wu, M.C. NanoPen: Dynamic, Low-power, and light-actuated patterning of nanoparticles. *Nano Lett.* **2009**, *9*, 2921–2925. [CrossRef]

70. Yang, S.-M.; Harishchandra, P.T.; Yu, T.-M.; Liu, M.-H.; Hsu, L.; Liu, C.-H. Concentration of magnetic beads utilizing light-induced electro-osmosis flow. *IEEE Trans. Magn.* **2011**, *47*, 2418–2421. [CrossRef]

71. Ota, S.; Wang, S.; Wang, Y.; Yin, X.; Zhang, X. Lipid bilayer-integrated optoelectronic tweezers for nanoparticle manipulations. *Nano Lett.* **2013**, *13*, 2766–2770. [CrossRef]

72. Liang, W.; Liu, L.; Lai, S.H.-S.; Wang, Y.; Lee, G.-B.; Li, W.J. Rapid assembly of gold nanoparticle-based microstructures using optically-induced electrokinetics. *Opt. Mater. Express* **2014**, *4*, 2368. [CrossRef]

73. Liu, N.; Liang, W.; Mai, J.D.; Liu, L.; Lee, G.-B.; Li, W.J. Rapid fabrication of nanomaterial electrodes using digitally controlled electrokinetics. *IEEE Trans. Nanotechnol.* **2014**, *13*, 245–253. [CrossRef]

74. Liang, W.; Liu, L.; Wang, Y.; Lee, G.-B.; Li, W.J. Rapid assembly of CARBON Nanoparticles into electrical elements by optically-induced electroosmotic flow. *IEEE Trans. Nanotechnol.* **2018**, *17*, 1045–1052. [CrossRef]

75. Liu, N.; Wei, F.; Liu, L.; Lai, H.S.S.; Yu, H.; Wang, Y.; Lee, G.-B.; Li, W.J. Optically-controlled digital electrodeposition of thin-film metals for fabrication of nano-devices. *Opt. Mater. Express* **2015**, *5*, 838. [CrossRef]

76. Liu, N.; Wang, F.; Liu, L.; Yu, H.; Xie, S.; Wang, J.; Wang, Y.; Lee, G.-B.; Li, W.J. Rapidly patterning micro/nano devices by directly assembling ions and nanomaterials. *Sci. Rep.* **2016**, *6*, 32106. [CrossRef]

77. Li, P.; Liu, N.; Yu, H.; Wang, F.; Liu, L.; Lee, G.-B.; Wang, Y.; Li, W.J. Silver nanostructures synthesis via optically induced electrochemical deposition. *Sci. Rep.* **2016**, *6*, 28035. [CrossRef]

78. Li, M.; Liu, N.; Li, P.; Shi, J.; Li, G.; Xi, N.; Wang, Y.; Liu, L. Performance investigation of multilayer MoS2 thin-film transistors fabricated via mask-free optically induced electrodeposition. *ACS Appl. Mater. Interfaces* **2017**, *9*, 8361–8370. [CrossRef]

79. Liu, N.; Li, M.; Liu, L.; Yang, Y.; Mai, J.; Pu, H.; Sun, Y.; Li, W.J. Single-step fabrication of electrodes with controlled nanostructured surface roughness using optically-induced electrodeposition. *J. Micromechan. Microeng.* **2018**, *28*, 025011. [CrossRef]

80. Wang, S.; Liang, W.; Dong, Z.; Lee, V.G.B.; Li, W.J. Fabrication of micrometer—and nanometer-scale polymer structures by visible light induced dielectrophoresis (DEP) force. *Micromachines* **2011**, *2*, 431–442. [CrossRef]

81. Liu, N.; Li, P.; Liu, L.; Yu, H.; Wang, Y.; Lee, G.-B.; Li, W.J. 3-D Non-UV digital printing of hydrogel microstructures by optically controlled digital electropolymerization. *J. Microelectrom. Syst.* **2015**, *24*, 2128–2135. [CrossRef]

82. Li, Y.; Lai, S.H.S.; Liu, N.; Zhang, G.; Liu, L.; Lee, G.-B.; Li, W.J. Fabrication of high-aspect-ratio 3D Hydrogel microstructures using optically induced electrokinetics. *Micromachines* **2016**, *7*, 65. [CrossRef] [PubMed]

83. Yang, W.; Yu, H.; Li, G.; Wang, Y.; Liu, L. Tissue engineering: High-throughput fabrication and modular assembly of 3D heterogeneous microscale tissues (small 5/2017). *Small* **2017**, *13*, 1–11. [CrossRef]

84. Li, P.; Yu, H.; Liu, N.; Wang, F.; Lee, G.-B.; Wang, Y.; Liu, L.; Li, W.J. Visible light induced electropolymerization of suspended hydrogel bioscaffolds in a microfluidic chip. *Biomater. Sci.* **2018**, *6*, 1371–1378. [CrossRef] [PubMed]

85. Liu, N.; Liang, W.; Liu, L.; Wang, Y.; Mai, J.D.; Lee, G.-B.; Li, W.J. Extracellular-controlled breast cancer cell formation and growth using non-UV patterned hydrogels via optically-induced electrokinetics. *Lab Chip* **2014**, *14*, 1367. [CrossRef] [PubMed]

86. Chan-Park, M.B.; Yan, Y.; Neo, W.K.; Zhou, W.; Zhang, J.; Yue, C.Y. Fabrication of high aspect ratio poly(ethylene glycol)-containing microstructures by UV embossing. *Langmuir* **2003**, *19*, 4371–4380. [CrossRef]

87. Terakawa, M.; Torres-Mapa, M.L.; Takami, A.; Heinemann, D.; Nedyalkov, N.N.; Nakajima, Y.; Hördt, A.; Ripken, T.; Heisterkamp, A. Femtosecond laser direct writing of metal microstructure in a stretchable poly(ethylene glycol) diacrylate (PEGDA) hydrogel. *Opt. Lett.* **2016**, *41*, 1392–1395. [CrossRef]

88. Zhao, Y.; Liang, W.; Zhang, G.; Mai, J.D.; Liu, L.; Lee, G.-B.; Li, W.J. Distinguishing cells by their first-order transient motion response under an optically induced dielectrophoretic force field. *Appl. Phys. Lett.* **2013**, *103*, 183702. [CrossRef]

89. Liang, W.; Zhao, Y.; Liu, L.; Wang, Y.; Li, W.J.; Lee, G.-B. Determination of cell membrane capacitance and conductance via optically induced electrokinetics. *Biophys. J.* **2017**, *113*, 1531–1539. [CrossRef]

90. Zhao, Y.; Lai, H.S.S.; Zhang, G.; Lee, G.-B.; Li, W.J. Measurement of single leukemia cell's density and mass using optically induced electric field in a microfluidics chip. *Biomicrofluidics* **2015**, *9*, 022406. [CrossRef]

91. Chiu, T.-K.; Chao, A.-C.; Chou, W.-P.; Liao, C.-J.; Wang, H.-M.; Chang, J.-H.; Chen, P.-H.; Wu, M.-H. Optically-induced-dielectrophoresis (ODEP)-based cell manipulation in a microfluidic system for high-purity isolation of integral circulating tumor cell (CTC) clusters based on their size characteristics. *Sens. Actuators B Chem.* **2018**, *258*, 1161–1173. [CrossRef]

92. Chu, P.-Y.; Liao, C.-J.; Hsieh, C.-H.; Wang, H.-M.; Chou, W.-P.; Chen, P.-H.; Wu, M.-H. Utilization of optically induced dielectrophoresis in a microfluidic system for sorting and isolation of cells with varied degree of viability: Demonstration of the sorting and isolation of drug-treated cancer cells with various degrees of anti-cancer drug resistance gene expression. *Sens. Actuators B Chem.* **2019**, *283*, 621–631. [CrossRef]

93. Huang, S.-H.; Hung, L.-Y.; Lee, G.-B. Continuous nucleus extraction by optically-induced cell lysis on a batch-type microfluidic platform. *Lab Chip* **2016**, *16*, 1447–1456. [CrossRef] [PubMed]

94. Hsiao, Y.-C.; Wang, C.-H.; Lee, W.-B.; Lee, G.-B. Automatic cell fusion via optically-induced dielectrophoresis and optically-induced locally-enhanced electric field on a microfluidic chip. *Biomicrofluidics* **2018**, *12*, 034108. [CrossRef]

95. Hwang, H.; Oh, Y.; Kim, J.-J.; Choi, W.; Park, J.-K.; Kim, S.-H.; Jang, J. Reduction of nonspecific surface-particle interactions in optoelectronic tweezers. *Appl. Phys. Lett.* **2008**, *92*, 024108. [CrossRef]

96. Valley, J.K.; Ohta, A.T.; Neale, S.L.; Jamshidi, A.; Wu, M.C.; Hsu, H.-Y. Optoelectronic tweezers as a tool for parallel single-cell manipulation and stimulation. *IEEE Trans. Biomed. Circuits Syst.* **2009**, *3*, 424–431. [CrossRef]

97. Lin, Y.-H.; Lee, G.-B. Optically induced flow cytometry for continuous microparticle counting and sorting. *Biosens. Bioelectron.* **2008**, *24*, 572–578. [CrossRef]

98. Lin, Y.-H.; Lee, G.-B. An integrated cell counting and continuous cell lysis device using an optically induced electric field. *Sens. Actuators B Chem.* **2010**, *145*, 854–860. [CrossRef]

99. Huang, K.-W.; Su, T.-W.; Ozcan, A.; Chiou, P.-Y. Optoelectronic tweezers integrated with lensfree holographic microscopy for wide-field interactive cell and particle manipulation on a chip. *Lab Chip* **2013**, *13*, 2278–2284. [CrossRef]

100. Lee, G.-B.; Chang, C.-J.; Wang, C.-H.; Lu, M.-Y.; Luo, Y.-Y. Continuous medium exchange and optically induced electroporation of cells in an integrated microfluidic system. *Microsyst. Nanoeng.* **2015**, *1*, 15007. [CrossRef]

101. Witte, C.; Kremer, C.; Cooper, J.M.; Neale, S.L. Continuous cell lysis in microfluidics through acoustic and optoelectronic tweezers. *SPIE MOEMS-MEMS* **2013**, *8615*, 86150. [CrossRef]
102. Witte, C.; Wilson, R.; Cooper, J.M.; Neale, S.L. OET Meets Acoustic Tweezing. In Proceedings of the Optical Trapping and Optical Micromanipulation IX, San Diego, CA, USA, 10 October 2012; pp. 411–418.

 *micromachines*

Review

Autonomous and In Situ Ocean Environmental Monitoring on Optofluidic Platform

Fang Wang [1,2], Jiaomeng Zhu [1,2], Longfei Chen [1,2], Yunfeng Zuo [1,2], Xuejia Hu [1,2] and Yi Yang [1,2,*]

1   Key Laboratory of Artificial Micro/Nano Structure of Ministry of Education, School of Physics and Technology, Wuhan University, Wuhan 430072, China; wangfang2200@whu.edu.cn (F.W.); zhujiaomeng@whu.edu.cn (J.Z.); coiyhm@whu.edu.cn (L.C.); zuoyf@whu.edu.cn (Y.Z.); huxuejiawuda@whu.edu.cn (X.H.)
2   Shenzhen Research Institute, Wuhan University, Shenzhen 518000, China
*   Correspondence: yangyiys@whu.edu.cn; Tel.: +86-(027)-6875-2989-8103

Received: 30 November 2019; Accepted: 7 January 2020; Published: 8 January 2020

Abstract: Determining the distributions and variations of chemical elements in oceans has significant meanings for understanding the biogeochemical cycles, evaluating seawater pollution, and forecasting the occurrence of marine disasters. The primary chemical parameters of ocean monitoring include nutrients, pH, dissolved oxygen (DO), and heavy metals. At present, ocean monitoring mainly relies on laboratory analysis, which is hindered in applications due to its large size, high power consumption, and low representative and time-sensitive detection results. By integrating photonics and microfluidics into one chip, optofluidics brings new opportunities to develop portable microsystems for ocean monitoring. Optofluidic platforms have advantages in respect of size, cost, timeliness, and parallel processing of samples compared with traditional instruments. This review describes the applications of optofluidic platforms on autonomous and in situ ocean environmental monitoring, with an emphasis on their principles, sensing properties, advantages, and disadvantages. Predictably, autonomous and in situ systems based on optofluidic platforms will have important applications in ocean environmental monitoring.

Keywords: optofluidics; ocean monitoring; colorimetric method

---

1. Introduction

The ocean is a vast repository of resources for human society and an essential material foundation for the sustainable development of society and environment. However, with continuous development and utilization of oceans by human beings, the ocean ecological environment has been gradually destroyed, and the marine ecological system has been severely damaged. Marine natural disasters, such as red tides and tsunamis, occurred frequently, bringing substantial economic losses and social impacts. Therefore, it is urgent to protect the ocean environment. Ocean environmental monitoring is an important link in protecting the marine environment. Developing the technology of ocean environmental monitoring is of great significance for the early predicting of marine disasters, preventing and reducing harmful disasters [1,2].

Recently, increasing attention has been paid to the autonomous and in situ ocean monitoring. The primary parameters of ocean environmental monitoring include nutrients (e.g., nitrate, phosphate, silicate), pH, DO, and heavy metals [3–9]. Traditional ocean environmental monitoring mainly relies on manual processing of a large amount of seawater samples and laboratory analysis, which consists of multiple steps (e.g., sampling, transport, pretreatment, instrument analysis) and is rather costly and time-consuming. Besides, the obtained results may be inaccurate since the seawater samples could undergo unexpected reactions during the long-time operation. The relatively low representative and timeliness make it challenging to meet the requirements of rapid detection of marine indicators.

Moreover, bulky and expensive high tech instruments and professional personnel are required for traditional marine monitoring. As a result, it is difficult to support ocean environmental monitoring for early warning and management.

Optofluidics is a relatively new technology field that synergistically integrates the optics and microfluidics; it provides many particular characteristics for enhancing the sensing performance and the minimization of systems [10–16]. Fluid is the natural carrier of various micro-nano particles (including chemical macromolecules and biomolecules). Together with the fluid, particles can directly flow into the optofluidic system, where chemical reactions can complete quickly. Optofluidics has the potential to easily detect various optical parameters with high sensitivity and accuracy, such as refractive index, fluorescence, and absorbance. It avoids the complicated processes in traditional detection methods. Besides, the most favored materials for optofluidic chips are polydimethylsiloxane (PDMS) [15] and polymethyl methacrylate (PMMA), both of them are cheap and easily replaceable. These advantages of high integration, sensitivity and accuracy, and low cost make the optofluidic technology widely used in environmental monitoring [5,17] and biochemical sensing.

The development of an autonomous and in situ ocean environmental monitoring system is of high priority in oceanographic research [18]. Optofluidics bring new ideas for the miniaturization of detection systems and has been increasingly considered as a powerful technology to realize ocean environmental monitoring. At present, although a few integrated systems based on optofluidic platform have been reported, a detailed review on its applications on ocean environmental monitoring is absent. Here, we describe the optofluidics-based autonomous monitoring of nutrients, pH, DO, and heavy metals in oceans, with a focus on their principles, sensing properties, advantages, and disadvantages compared with other reported ocean monitoring methods. Much attention has been paid to the most commonly used colorimetric/spectrophotometric detection method, as it has the capability of integrating all of the processes involved in the analysis into a small chip. At last, we forecast that an autonomous and in situ system based on optofluidic platforms will play important roles in the development of ocean environmental monitoring.

2. Nutrients

Soluble inorganic nitrogen, phosphorus, and silicate in seawater are essential nutrients for the survival of marine organisms. A moderate amount of nutrients in seawater promotes the growth of biology and microorganisms, while inadequate nutrients restrict the growth of phytoplankton and excessive nutrients are prone to cause eutrophication and even further lead to harmful algal blooms, extreme depletion of DO, and even death of aquatic organism [19–21]. Generally, the nitrite concentration in seawater is very low and stable. A too high concentration of nitrite or dramatic changes of nitrite concentration often indicate changes in the ocean environment. Phosphate is an important indicator of eutrophication, and the distribution of silicate affects the community structure of planktonic algae. Accurate quantification of such nutrients in oceans is essential for comprehending the dynamics of marine ecosystems and forecasting the occurrence of harmful red tides [5].

Various techniques, such as colorimetry [22–34], chemiluminescence [35,36], fluorimetry [37–40], electrochemistry [41–43], and chromatography [44–46], have been proposed for nutrient determination. Among them, the colorimetric method using chromogenic agents is one of the most favored detection approaches due to its stability, excellent detection limits, simplicity, high cost efficiency, and analytical feasibility. It operates by adding a color-developing agent to samples, and the absorbance spectrum of the product solution is proportional to the concentration of the nutrient based on the Beer–Lambert law:

$$A(\lambda) = -\lg\left(\frac{I}{I_0}\right) = \varepsilon(\lambda)cl \tag{1}$$

where, $A(\lambda)$ is the absorbance of the solution at a wavelength λ, I_0 is the intensity of the initial monochromatic light, I is the transmitted intensity of monochromatic light, $\varepsilon(\lambda)$ is the molar absorption coefficient, c is the concentration of the analyte, and l is the length of light path. To improve the reliability and reduce cost, the discrete and auxiliary optical devices, such as light-emitting diodes (LEDs) and photodetectors, are usually used to construct the autonomous detection systems.

2.1. Nitrate and Nitrite

Among various colorimetric assay chemistries, the Griess assay method has been the mainstay for nitrite analysis for over a century [22–34]. The mechanism of the Griess assay method is that under acidic conditions, nitrite reacts with sulfanilic acid to produce a diazonium salt, which is then coupled to *N*-(1-naphthyl) ethylenediamine (NED), resulting in pink azo compounds that can be detected at 543 nm. For nitrate determination, nitrate should be reduced to a more reactive nitrite by copperized cadmium [6,23,47], zinc [25], vanadium chloride [7,48], etc.

Figure 1 shows several examples of optofluidic chips and assembled systems using the Griess method for nitrate and nitrite detection. Beaton et al. firstly reported a new generation of miniaturized, in situ, and stand-alone systems with sufficient stability as well as analytical performance for the determination of nitrate and nitrite in natural waters based on optofluidics [33]. As shown in Figure 1a–c, the optofluidic platform consists of a tinted PMMA substrate (5.0 mm thick) micromilled with three absorption cells, three pairs of green LEDs (525 nm), and high sensitivity photodiodes for the detection of nitrate/nitrite with different concentrations. Dark PMMA was used for reducing the background light of LED reaching the photodiode. The optofluidic platform integrated with a custom electronics package for the operational control, data collection, analysis, and transfer. The whole monitoring system has a small size (100 mm diameter and 200 mm height), and low power consumption (about 1.5 W). Autonomous and in situ determination of nitrate and nitrite was deployed with this integrated system, the linear range is up to 350 µM, and the limit of detections (LODs) are as low as 0.025 µM and 0.02 µM for nitrate and nitrite, respectively, making it suitable to be applied in almost all natural waters.

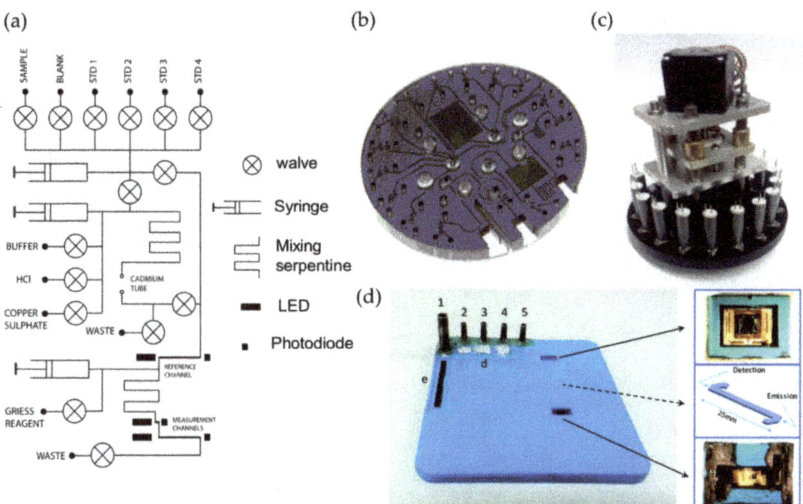

Figure 1. The optofluidic platforms and integrated devices used for the determination of nitrate and nitrite. (**a–c**) are the flow diagram, the corresponding photograph, and autonomous integrated device of the microanalyzer used to determine the total nitrate and nitrite. (**d**) The optofluidic chip and the photographs of the autonomous integrated device designed by Cisneros et al. Images reproduced from [6,33].

Although the nitrate and nitrite detection systems based on optofluidic technology have been successfully applied in marine online monitoring, there are some drawbacks to be overcome.

For example, optofluidic chips are prone to having the problems of bubble formation and blockages due to the narrow microchannels. Martinez-Cisneros et al. developed a compact and automated lab-on-a-chip (LOC) device that integrated microfluidic platform, a highly sensitive colorimetric detection module, and electronics platform for online determination of nitrate and nitrite, as shown in Figure 1d [6]. An embedded and monolithic microcolumn, which was internally installed with copperized cadmium granules, was used as a sample pretreatment module for the reduction of nitrate to nitrite. The device also integrated a bubble removal structure to minimize blockages and avoid potential interferences caused by bubbles, and was successfully applied to the continual monitoring of nitrate and nitrite.

All of the above systems for nitrate detection employ the indirect method, i.e., the concentration of background nitrite is firstly obtained directly after its reaction with the chromogenic agents, the nitrate is then reduced to nitrite, and the second measurement of the total nitrate and nitrite is conducted. The concentration of nitrate finally could be obtained by subtraction. This indirect and time-consuming method adds complexity to the monitoring system. Cogan et al. developed a highly integrated system for nitrate determination. The system was based on the direct and simplified chromotropic acid reagent method, which eliminates the reduction step of nitrate to nitrite [30], as shown in Figure 2. The principle is that the chromotropic acid reacts with nitrate ions in a sulphuric acid medium, resulting in yellow products that can be detected at 430 nm. With this method, a detection range for nitrate from 0.9 to 80 mg/L was achieved with a LOD of 0.73 mg/L. The advantages of simplicity, low cost, low reagent consumption, compact design, and high sample throughput make it an ideal candidate for applying in in situ detection of nitrate.

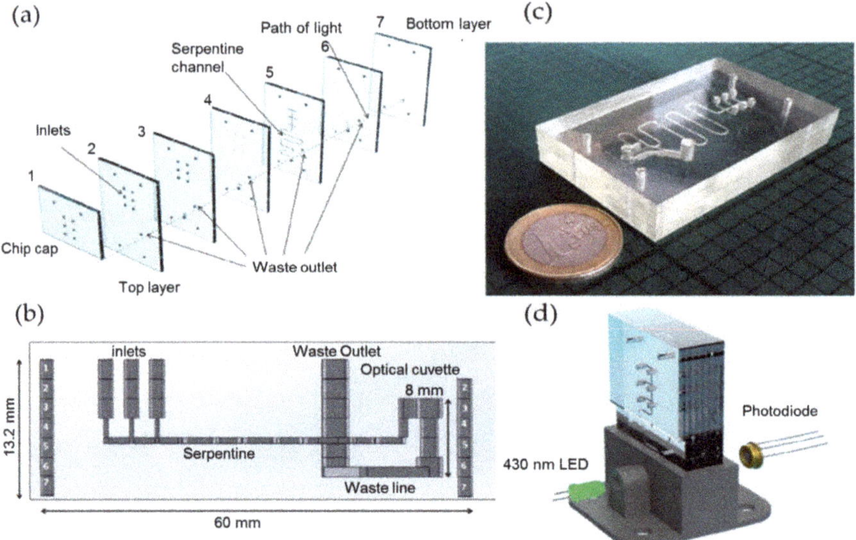

Figure 2. (**a**) Assembly diagram of microchip layers. (**b**) The chip layers that are fully assembled and the micro-cuvette. (**c**) The PDMS microfluidic chip. (**d**) Setup of the microfluidic chip and the optical detection module with LED and photodiode. Images reproduced from [30].

The traditional colorimetric analysis for nitrate and nitrite detection is time-consuming as it requires determining the calibration curve of the system, which means that a couple of standard samples need to be prepared and detected firstly. To overcome this drawback, some modified approaches based on the Griess assay have been proposed. Recently, Shi et al. demonstrated a robust differential colorimetric method for nitrite detection [24], as shown in Figure 3. The nitrite samples and Griess reagent were pumped into the designed tree-like network on a microchip made from PDMS, forming a unique concentration profile as well as a color gradient network, which contained sufficient information for the determination of nitrite. Only one sample is required for this differential colorimetric method, leading to less consumption of both time and power. The measuring errors for the used two nitrite solutions (0.50 mM and 0.33 mM) are 1.16% and 0.50%, respectively. Compared with the classic method that required a calibration curve, the stability and accuracy of the system are improved by approximately ten times and six times, respectively.

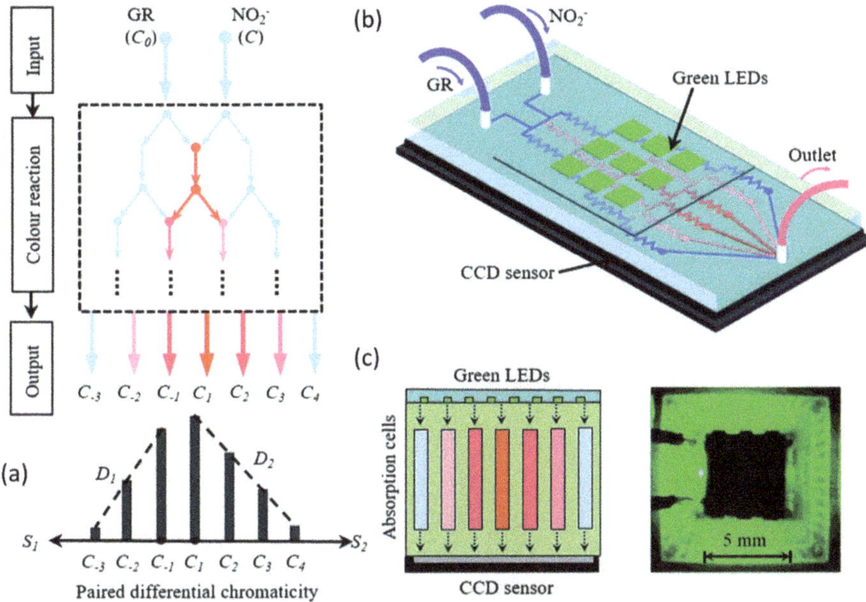

Figure 3. (**a**) Schematic diagram for forming the bidirectional differential concentration. (**b**) Optofluidic platform setup. (**c**) Left: optical detection system with green LEDs and CCD image sensor. Right: physical diagram of the integrated system. Images reproduced from [24].

2.2. Phosphate

A variety of approaches have been used for the determination of phosphate in seawater, including colorimetric detection, fluorescent detection, and electrochemical detection. All of these are reagent-based methods, as phosphate cannot be detected directly. Generally, autonomous systems tend to employ the colorimetric method rather than fluorescent or electrochemical methods [49,50].

Legiret et al. reported a high-performance optofluidic system for marine phosphate determination using the vanadomolybdate method [51]. The basic principle of this method is that under acidic conditions, orthophosphate reacts with vanadium polymolymolybdate reagent directly and rapidly, resulting in a stable yellow product that can be detected at 375 nm [52,53]. Figure 4a–c shows the optofluidic chip and the assembled system with this approach. The system incorporated an optofluidic chip in tinted PMMA, an optical detection module with high power UV-LEDs as a light source, and photodiodes as absorbance detectors, an embedded control electronics and syringe pumps. The integrated optofluidic analyzer has a small physical size of 22 cm (height) by 10 cm (diameter). Experimental results showed that it has a wide linear range from 0.1 μM to 60 μM and a LOD of 52 nM for phosphate detection. This integrated microanalyzer features high measurement accuracy and resolution, while requires low power and reagent consumption, and it has been successfully deployed in autonomous and in situ phosphate monitoring.

Among all reagent based colorimetric approaches for the determination of phosphate, phosphomolybdenum blue method is the most widely used one [5,51–56]. Its principle is that under acidic conditions, orthophosphate reacts with ammonium molybdate, producing a light yellow ammonium phosphomolybdate, which is then reduced by the reducing agent (e.g., ascorbic acid) into a blue compound with strong chromogenic capacity—'molybdenum blue'. Duffy et al. reported a portable, compact lab-on-a-disc device based on the phosphomolybdenum blue method for in situ quantitation of water phosphate [52]. As shown in Figure 4d,e, the integration of a microfluidic disc allows the use of an optical path length as long as 75 mm for improving sensitivity. The device also integrated a low-cost optical detection system formed by a pair of LEDs and photodiodes. The total mass and the physical size of the device is 2 kg and 20 cm × 18 cm × 14 cm, respectively. The experimental results showed that the detection range for phosphate is 14–800 µg/L, and the LOD is as low as 5 µg/L. Grand et al. reported a newly developed LOC analyzer for long-term and in situ phosphate monitoring [56]. To obtain the best analytical sensitivity, the influence of sample temperature and salinity on the reliability and accuracy was studied, and the reaction parameters were evaluated and optimized. The analyzer is a viable candidate to be deployed in ocean phosphate monitoring as it owns many merits such as perfect stability and robustness, small size, simple to operate, and low energy consumption (1.8 W). It also features a LOD of 30 nM and a wide linear range from 0.05 to 13 µM, with a precision less than 10%.

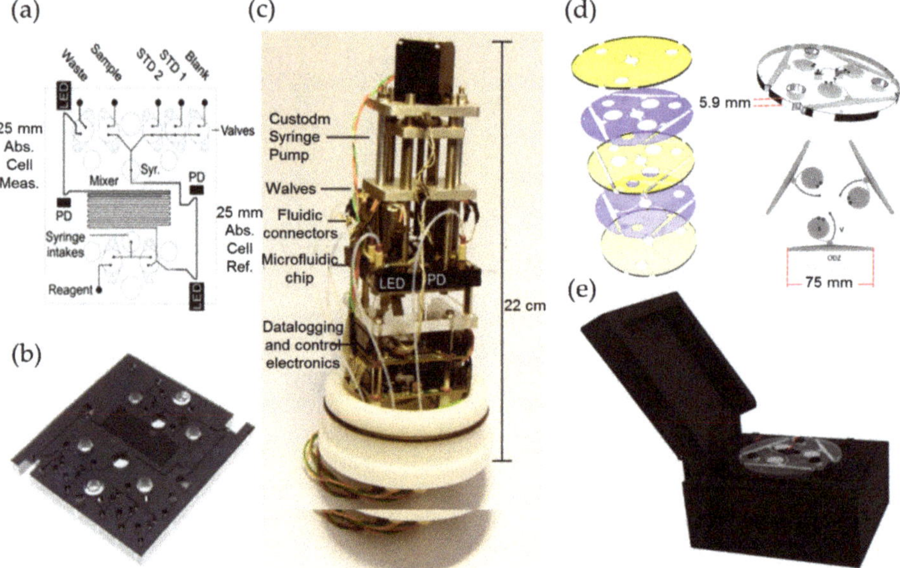

Figure 4. The optofluidic chips and assembled system used for phosphate detection. (**a–c**) Flow diagrams, the corresponding photographs, and the autonomous integrated devices of the microanalyzers used to determine the total phosphate ion concentrations. (**d,e**) The rendered images showing each layer of the microfluidic disc and the phosphate sense system. Images reproduced from [51,52].

For many optofluidic autonomous monitoring systems based on the colorimetric method, the LOD is not comparable to the results obtained by traditional methods because of the limitations on the size of the absorption cell on microchips. To enhance the phosphate absorption and obtain better measurement accuracy, Zhu et al. proposed an innovative on-chip Fabry–Pérot microcavity, which consists of two parallel reflectors made by a pair of fiber facets coated with Au films (Figure 5) [5]. Light is reflected many times in the microcavity, thus lengthening the optical path and enhancing absorption. The length of the absorption cell is shortened from several centimeters to 300 μm, and the device still features a LOD as low as 0.1 μM for phosphate detection. Combining the specially designed passive microreactor with semilunar barriers for rapid and enough chromogenic reaction, the required time for detection is shortened from several minutes to six seconds.

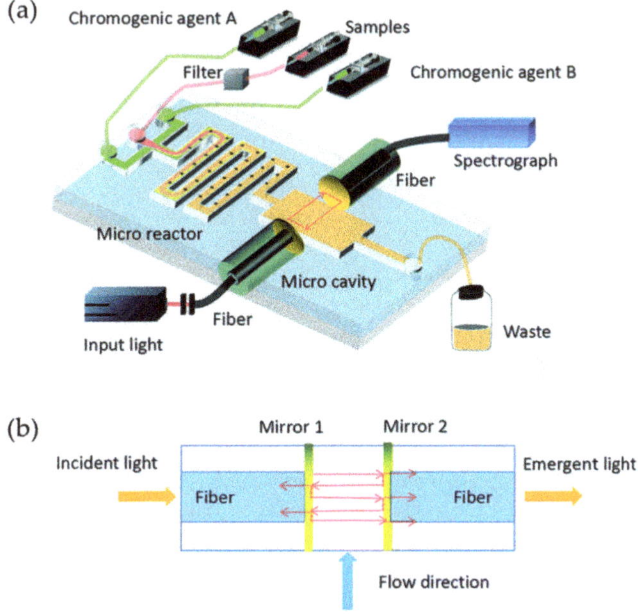

Figure 5. A Fabry–Pérot resonator-based optofluidic device used for monitoring of phosphate in oceans. (**a**) Setup of the optofluidic device. Microchannels with crescent batteries are designed for rapid and enough mixing of regents. (**b**) The microcavity is formed by a pair of optical fibers, on which surfaces are coated with gold. Images reproduced from [5].

2.3. Silicate

Silicomolybdenum yellow method and silicomolybdenum blue method are the two standard methods that are commonly used for detection of silicate in seawater. The former method is rapid but features poor sensitivity and is easily interfered with by salinity, making it unsuitable for low concentration analysis. The silicomolybdenum blue method is widely used for the determination of marine silicate as it has the advantage of high sensitivity. Its mechanism is based on the reaction of silicate with ammonium molybdate to form a yellow silicomolybdate complex, which is further reduced to a stable and detectable blue silicomolybdenum product by ascorbic acid. Cao et al. reported a new generation of optofluidic sensor that has been successfully applied for analyzing the silicate in seawater [57]. The silicate sensor contained four functional modules, including the fluid driving system, a microchip made of tinted PMMA for the mixing and reaction of reagents, a control circuit, and a sensor peripheral. Three absorbance cells were designed for a wide detection range, and three pairs of 810 nm LEDs and photodiodes were used accordingly to perform spectrophotometry analysis. The LOD of the sensor was as low as 45.1 nM, and the linear determination range of the sensor was 0 to 400 μM. An offshore experiment proved that this optofluidic sensor has the advantages of low regent consumption, high accuracy, and robustness.

3. pH

Human activities produce a large amount of carbon dioxide (CO_2). CO_2 released into the atmosphere could be absorbed by the ocean, leading to significant chemical changes like ocean acidification. The absorption of CO_2 also regulates the pH of seawater, which is a key parameter for the aquatic organism and influences the ecosystems and biogeochemical cycles of ocean [8,58–62]. A water environment with pH from 6.5 to 8.0 is required for aquatic life, the quantity of water-based life reduces outside of this range because the physiological systems that organisms lived by are affected. Toxic heavy metals (Cd, Pb, Hg, etc.) become more soluble at low pH, thus increasing the toxicity levels of the living environment for organisms [62]. As a result, accurate and timely determination of pH is dispensable for fully understanding the marine carbon cycle and changes of the ecosystem [4,63,64].

Glass electrode method is one of the most popular methods used for pH measurement; accurate results could be achieved if the glass electrodes are maintained and operated properly. However, the glass electrode method also has several drawbacks, such as physical fragility, leakage of the reference electrode buffer, and various responses of the glass electrode with salinity and temperature. All of these disadvantages limit its applications for long-term ocean monitoring. The principle of the most commonly used colorimetric detection of pH is adding indicator dyes, such as thymol blue [65,66], phenol red [67], cresol red [68], and meta-cresol purple [69] to seawater samples, resulting in colored compounds that indicate the pH value.

There are a few examples of optofluidic devices for in situ colorimetric pH measurements. Florea et al. developed a low cost and accurate optofluidic device for pH determination based on polyaniline (PAni), the detectable range for this device is from pH 2 to 12 [70]. The device integrated PAni-based coatings and spectrophotometry method to measure pH. The absorbance spectrum of PAni coatings changes when solutions at various pH values flow along the microchannel. This microfluidic sensor requires no more indicator reagent, thus reducing the complexity of pH detection. Besides, this work evaluated the feasibility of using a digital color camera (e.g., mobile phone with integrated digital cameras) instead of a spectrophotometer to perform colorimetric analysis, which will dramatically extend its applications [71,72]. However, the system is not fully autonomous and requires manual input, like taking photographs and sampling.

More fully developed autonomous systems for seawater pH measurement have been reported. Rérolle et al. reported a colorimetric pH sensor based on optofluidics for autonomous ocean monitoring. The pH indicator dye of thymol blue was employed in this case [66]. The optofluidic chip is made of tinted PMMA, it consists of a static mixer formed by a serpentine channel as long as 2.2 m, and an absorption cell with a size of 700 μm (length) by 700 μm (width). Its principle is that samples and indicator dye solutions were effectively mixed in the static mixer, then the mixer flows into the absorption cell for optical detection. The detection module consists of a light source with a tricolored LED and a detector with a linear array photodiode spectrometer. The system was successfully implemented in shipboard deployment and a precision of 0.001 pH unit for short-term pH monitoring could be obtained. Lai et al. designed an autonomous optofluidic chemical analyzer for both pH and partial pressure of CO_2 (pCO_2) [17]. A pH indicator was used for the determination of the pCO_2. The indicator was pumped into the gas permeable membrane and then given an optical response when reacting with pCO_2. With the beam-splitter design shown in Figure 6, the analyzer has a precision of $\sim \pm 0.5$ μatm for pCO2 and $\sim \pm 0.0005$ pH for pH detection. With the specially designed regent bag and adequate power supply, over 8500 measurements could be obtained for every scuba diving. However, the primary defect of this optofluidic analyzer is that it is more complicated compared to other pH and pCO_2 sensors, such as the single-ended electrode or optode sensor.

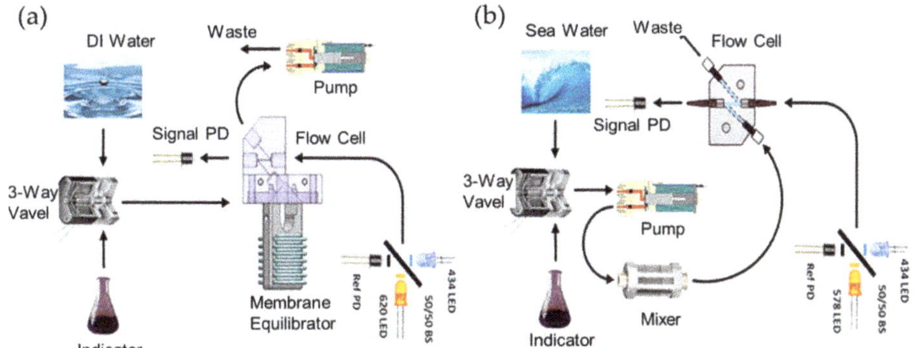

Figure 6. (**a**) Schematics of the detection system for SAMI-CO_2. (**b**) Schematics of the detection system for SAMI-pH. Images reproduced from [17].

4. DO

Dissolved oxygen is an indispensable substance for the aquatic organism and a vital parameter to characterize the metabolism of marine ecosystems [73–75]. The DO content in seawater is influenced by biochemical and physiological activities. As the DO content of polluted seawater is lower than that of natural seawater, the determination of DO also helps to evaluate the hypoxia and pollution in marine environments.

Many methods have been used to detect DO in oceans, including the iodometric titration method, which is internationally recognized as a benchmark [76], the most widely used electrochemical method, and the optical method, which is mainly based on the principle of fluorescence quenching. Each method has its own merits and drawbacks. The iodometric titration determination has the advantage of high accuracy, but it features cumbersome detection procedures and is not suitable for continuous online detection [77]. Electrochemical method characterizes for rapid detection and simple operation, but it has limitations such as the requirements for calibration and regular maintaining, and the electrode is easily poisoned. The optical DO sensor based on fluorescence quenching has the advantages of versatility and high sensitivity, as well as low toxicity, but the detection accuracy is still needed to improve [78].

Innovative research on the applications of optofluidics in DO detection have been reported. Mahoney et al. proposed a multilayer optofluidic device based on measuring the fluorescence quenching in a Ruthenium-based oxygen sensitive dye. Enhanced sensitivity was achieved by employing total internal reflection (TIR) of the excitation light. The optofluidic sensor exhibits high sensitivity for the detection of 0–20 ppm DO in water [79]. However, the automation and miniaturization of DO sensors for ocean monitoring based on optofluidic technology remain to be developed [80,81].

5. Heavy Metals

It is of great importance to monitor heavy metals in oceans as they can seriously affect the environment and human health. Generally, heavy metal refers to metals with specific gravities greater than 5 g/cm^3. It includes essential metals that are indispensable for the normal physiological activities of organic life such as iron (Fe); copper (Cu); magnesium (Mg); selenium (Se); zinc (Zn); manganese (Mn); nonessential metals including cadmium (Cd), mercury (Hg), silver (Ag), lead (Pb), and gold (Au); and some unusual metals with high atomic weight [82]. With the accumulation of essential metals in organisms, toxic side effects will occur if its concentration exceeds a certain threshold [83]. Nonessential metals could be hypertoxic even at a trace level, although they do not participate in the metabolic activities of organisms.

Many analytical techniques have been proposed for the determination of heavy metals in seawater, for example, the inductively coupled plasma mass spectroscopy (ICP-MS) method and atomic absorption spectrometry method. However, both of them require tedious sample preparation and pretreatment, expensive equipment, and professional personnel, making them insuitable in applications for autonomous, in situ, and continuous monitoring of heavy metals. In this case, ocean monitoring sensors based on colorimetric, fluorescent, and chemiluminescent methods appear as promising technologies due to the high sensitivity and feasibility, versatility, and reproducibility [84–95], and all of these methods have been truly applied for online analysis of heavy metals in seawater.

The colorimetric method for heavy metal monitoring is also based on Lambert–Beer's law. The principle is that under certain conditions, heavy metal ions react with a specific reagent, producing a new colored chemical solution, and the absorbance of the solution is related to the concentration of the heavy metal. Different heavy metals need different color developing agents when employing the colorimetric method. For example, silver salt is used for arsenic determination [96], dithizone is generally used for lead and zinc determination [97], while dimethylglyoxime is generally applied for nickel determination [98]. Milani et al. developed an autonomous LOC colorimetric system with high performance and low-cost for the in situ monitoring of dissolved Fe(II) and Mn in natural water [88]. The optofluidic platform consists of a microfluidic chip made of PMMA and two pairs of LEDs and photodiode detectors for detection of Fe(II) and Mn, as shown in Figure 7a,b. With this portable device, 12 samples of Fe(II) and 6 samples of Mn could be analyzed per hour with LODs of 27 nM and 28 nM, and precisions of 2.1% and 2.4% ($n = 19$), respectively. Lace et al. reported an innovative optofluidics system for monitoring of arsenic in water with the leucomalachite green dye method [91]. The principle involves the reactions of arsenic and potassium iodate under acidic conditions, and leucomalachite green is oxidized to malachite green by the liberated iodine, producing a green color with an absorption peak at 617 nm. Chromogenic agents and water samples are mixed and reacted in a microfluidic chip, as shown in Figure 7c. The leucomalachite green method was optimiszd and water arsenic within the range of 0.3–2 mg /L could be detected, the LOD of the system was 0.32 mg/L, and the average relative standard deviation (RSD) was 21.1%.

The fluorescent method is an excellent option for autonomous and in situ ocean heavy metal detection. Compared with the colorimetric method, higher sensitivity and lower LOD could be achieved with the fluorescent method. The basic mechanism of the fluorescent method is that the fluorescence parameters (such as fluorescence intensity, lifetime, and spectrum) change in response to the concentration variations of some ions. Until now, numerous fluorescent sensors based on chelation-enhanced fluorescence [98,99] photo-induced electron transfer [92], aggregation-induced emission effect [100], and intramolecular charge transfer [101] have already been reported for heavy metal monitoring. Leray's group developed several optofluidic devices that incorporated fluorimetric detection for the monitoring of heavy metals such as aluminum and cadmium [92,93,102]. Recently, the aluminum-sensing mechanism by PSSA–4-propoxysulfonate salicylaldehyde azine was proposed [92]. As shown in Figure 7e–g, the sensing system includes a microfluidic chip made from PDMS/glass, and a fluorescence detection platform consists of a LED (365) nm and a photomultiplier (PMT) detector. In brief, the PSSA and Al(III) are injected and mixed in a Y-shaped junction, the complexation of the ligand with the analyte occurs upon mixing, producing a fluorescence signal, which is collected and detected by the PMT. Regularly monodispersed and spaced microdroplets are formed, and a highly reproducible and reliable monitoring environment was provided as the system to adopt the microfluidic droplet technology. The experimental results showed that the system has a LOD of 153 nM. The newly designed fluorescent sensor for cadmium detection adopts a commonly used water-soluble commercial dye Rhod-5N, whose fluorescence intensity increased with the existence of toxic cadmium. The optofluidic device is made from PDMS chip and a glass substrate, as shown in Figure 7d. The famous staggered herringbone structure [98,102] was used to form a passive mixer for the thorough mixing of the Rhod-5N reactant and the Cd^{2+} analyte. The optofluidic device obtained a promising LOD of 0.45 µg/L in 3-(N-morpholino) propanesulfonic acid buffer at pH 7.

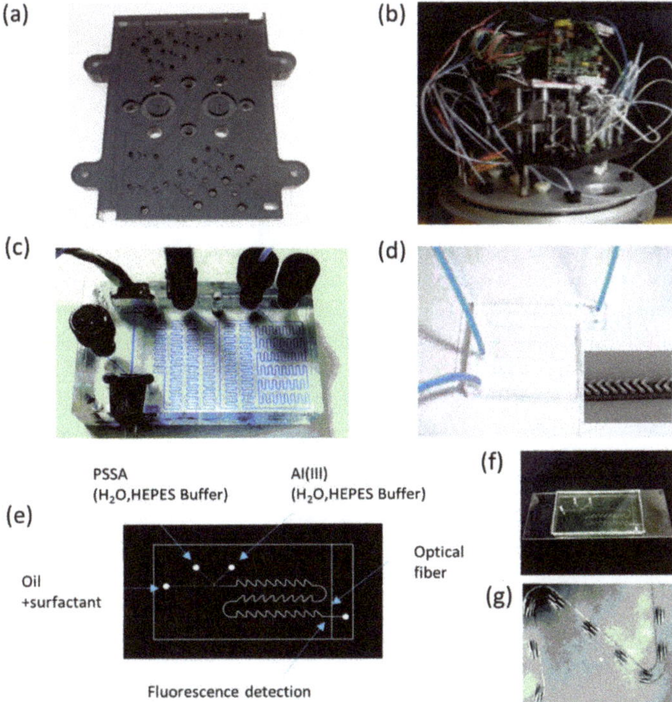

Figure 7. Spectrophotometric and fluorescent detection of heavy metal by optofluidic systems. (**a,b**) The photograph and the autonomous integrated device of the microanalyzers used to determine dissolved Fe and Mn with a spectrophotometric method. (**c**) The PMMA optofluidic chip used for arsenic detection with a colorimetric method. (**d**) The photograph of the optofluidic chip for fluorimetric detection of cadmium. (**e**) The scheme of the optofluidic set-up following the droplets approach for fluorescent detection of water-soluble aluminium. (**f,g**) The photographs of the microchip and the forming droplet of the fluorescent sensor. Images reproduced from [88,91,92,102].

The chemiluminescent method for quantitative analysis of heavy metals is mainly based on the linear relationship between the concentration of the analyte and the chemiluminescence intensity of the system under certain conditions. The most commonly used chemiluminescent reagents include luminol, acridine esters, and 1,2-dioxygen cycloethane. Based on the chemiluminescent reaction of Mn contained in the seawater sample with a luminol-based reagent, Christophe et al. developed an integrated in situ analyzer for detecting Mn (IISA-Mn) in the environment of the deep sea [89]. A miniaturized PDMS device with Tesla mixer structures was used to enhance the mixing and reaction efficiency, and a PMT detector was used to measure the chemiluminescent intensity. As a result of the integration of fluidic elements such as mixers, valves, and flow regulators, this analyzer based on optofluidics occupies less volume (3 L) and consumes less reagent (<130 mL/24 h) compared with its macroworld counterparts. The IISA-Mn gives a linear range of 0 to 500 nM and a LOD of 280 nM in seawater, making it sufficient for detecting geochemical anomalies such as hydrothermal ore deposition surveys. The system has been proven to be able to work flawlessly and continuously during the 8 h of an actual remotely operated vehicle dive.

6. Discuss and Outlook

There is more and more research on the optofluidic devices as optofluidic technology has advantages of low cost, small size, low reagent, and power demands. Although we emphasize the applications of optofluidic platforms for the monitoring of chemical parameters in ocean environment in this review, they could also find uses in biological monitoring in the marine ecosystem. For example, a variety of in situ bio/biochemical analyzers (e.g., adenosine triphosphate analyzer and gene analyzer) have been developed on optofluidic platforms, most of them have been evaluated in the real ocean environments including the deep sea [103,104].

As marine environmental pollution is increasingly severe, there is a growing demand for the further development of autonomous and in situ ocean monitoring. To realize continuously offshore monitoring with high practicability, stability, and credibility, smaller and lighter portable instruments with less power consumption and higher accuracy are needed. Optofluidic technology is one of the most promising approaches to realize the miniaturization and automation of ocean monitoring. This paper reviewed the optofluidic devices integrated with microelectromechanical systems that are applied for in situ ocean monitoring; up-to-date developments of optofluidics for autonomous ocean environmental monitoring were covered. Optofluidics has the challenge of ensuring that the overall system can achieve the required sensitivity for ocean applications where analytes exist at only trace levels. However, we firmly believe that based on the development trend of full automation and miniaturization of marine monitoring equipment, and with the development of the supported technology such as three-dimensional printing, inkjet printing, power-free pumping and micromachining technologies, the low-cost lab-on-chip systems based on optofluidic technology will inspire a promising future in the field of ocean observations.

Author Contributions: F.W., Y.Y., and J.Z. proposed the original idea, and planned the configuration and key topics. F.W. wrote the manuscript. J.Z., L.C., Y.Z. and X.H. revised the paper for language and quality. All authors took part in regular discussions and were involved in the completion of the manuscript. All authors have read and agreed to the published version of the manuscript.

Funding: This work was funded by open financial grants from Qingdao National Laboratory for Marine Science and Technology (No. QNLM2016ORP0410), National Natural Science Foundation of China (No. 11774274), National Key R&D Program of China (2018YFC1003200), and Foundation Research Fund of Shenzhen Science and Technology Program (No. JCYJ201708181112939064).

Conflicts of Interest: The authors declare no conflict of interest.

References

1. Anthony, K.R.; Kline, D.I.; Diaz-Pulido, G.; Dove, S.; Hoegh-Guldberg, O. Ocean acidification causes bleaching and productivity loss in coral reef builders. *Proc. Natl. Acad. Sci. USA* **2008**, *105*, 17442–17446. [CrossRef] [PubMed]

2. Nicholson, D.P.; Michel, A.P.M.; Wankel, S.D.; Manganini, K.; Sugrue, R.A.; Sandwith, Z.O.; Monk, S.A. Rapid mapping of dissolved methane and carbon dioxide in coastal ecosystems using the ChemYak Autonomous surface vehicle. *Environ. Sci. Technol.* **2018**, *52*, 13314–13324. [CrossRef] [PubMed]

3. Nightingale, A.M.; Beaton, A.D.; Mowlem, M.C. Trends in Microfluidic Systems for In Situ Chemical Analysis of Natural Waters. *Sens. Actuators B* **2015**, *221*, 1398–1405. [CrossRef]

4. Fukuba, T.; Imhof, A.; Matsunaga, A.; Takagi, N.; Yamamoto, T.; Okamura, K.; Naganuma, T.; Fujii, T. Development of miniaturized in situ analysis devices for biological and chemical oceanograph. In Proceedings of the 2005 3rd IEEE/EMBS Special Topic Conference on Microtechnology in Medicine and Biology, Oahu, HI, USA, 12–15 May 2005; pp. 56–59.

5. Zhu, J.M.; Shi, Y.; Zhu, X.Q.; Yang, Y.; Jiang, F.H.; Sun, C.J.; Zhao, W.H.; Han, X.T. Optofluidic marine phosphate detection with enhanced absorption using a Fabry–Pérot resonator. *Lab Chip* **2017**, *17*, 4025–4030. [CrossRef] [PubMed]

6. Martinez-Cisneros, C.; Rocha, Z.D.; Seabra, A.; Valdés, F.; Alonso-Chamarro, J. Highly integrated autonomous lab-on-a-chip device for on-line and in situ determination of environmental chemical parameters. *Lab Chip* **2018**, *18*, 1884–1890. [CrossRef] [PubMed]

7. Fang, T.Y.; Li, P.C.; Lin, K.N.; Chen, N.W.; Jiang, Y.Y.; Chen, J.X.; Yuan, D.X.; Ma, J. Simultaneous underway analysis of nitrate and nitrite in estuarine and coastal waters using an automated integrated syringe-pump-based environmental-water analyser. *Anal. Chim. Acta* **2019**, *1076*, 100–109. [CrossRef]

8. Clarke, J.S.; Achterberg, E.P.; Connelly, D.P. Developments in marine pCO_2 measurement technology; towards sustained in situ observations. *TrAC, Trends Anal. Chem.* **2017**, *88*, 53–61. [CrossRef]

9. Gumpu, M.B.; Sethuraman, S.; Krishnan, U.M. A review on detection of heavy metal ions in water – An electrochemical approach. *Sens. Actuators B* **2015**, *213*, 515–533. [CrossRef]

10. Psaltis, D.; Quake, S.R.; Yang, C. Developing optofluidic technology through the fusion of microfluidics and optics. *Nature* **2006**, *442*, 381–386. [CrossRef]

11. Monat, C.; Domachuk, P.; Eggleton, B.J. Integrated optofluidics: A new river of light. *Nat. Photonics* **2007**, *1*, 106–114. [CrossRef]

12. Fan, X.D.; White, I.M. Optofluidic microsystems for chemical and biological analysis. *Nat. Photonics* **2011**, *5*, 591–597. [CrossRef] [PubMed]

13. Zhang, Y.N.; Zhao, Y.; Zhou, T.; Wu, Q. Applications and developments of on-chip biochemical sensors based on optofluidic photonic crystal cavities. *Lab Chip* **2018**, *18*, 57–74. [CrossRef] [PubMed]

14. Zhu, J.M.; Zhu, X.Q.; Zuo, Y.F.; Hu, X.J.; Shi, Y.; Liang, L.; Yang, Y. Optofluidics: The interaction between light and flowing liquids in integrated devices. *Opto-Electron. Adv.* **2019**, *2*, 190007. [CrossRef]

15. Fujii, T. PDMS-based microfluidic devices for biomedical applications. *Microelectron. Eng.* **2002**, *61–62*, 907–914. [CrossRef]

16. Persichetti, G.; Grimaldi, I.A.; Testa, G.; Bernini, R. Multifunctional optofluidic lab-on-chip platform for Raman and fluorescence spectroscopic microfluidic analysis. *Lab Chip* **2017**, *17*, 2631–2639. [CrossRef]

17. Lai, C.Z.; DeGrandpre, M.D.; Darlington, R.C. Autonomous Optofluidic Chemical Analysers for Marine Applications: Insights from the Submersible Autonomous Moored Instruments (SAMI) for pH and pCO2. *Front. Mar. Sci.* **2018**, *4*, 1–11. [CrossRef]

18. Prien, R.D. The future of chemical in situ sensors. *Mar. Chem.* **2007**, *107*, 422–432. [CrossRef]

19. Robidart, J.; Magasin, J.D.; Shilova, I.N.; Turk-Kubo, K.A.; Wilson, S.T.; Karl, D.M.; Scholin, C.A.; Zehr, J.P. Effects of nutrient enrichment on surface microbial community gene expression in the oligotrophic North Pacific Subtropical Gyre. *ISME J.* **2019**, *13*, 374–387. [CrossRef]

20. Bricker, S.B.; Ferreira, J.G.; Simas, T. An integrated methodology for assessment of estuarine trophic status. *Ecol. Modell.* **2003**, *169*, 39–60. [CrossRef]

21. Perez-Ruzafa, A.; Campillo, S.; Fernandez-Palacios, J.M.; Garcialacunza, A.; Garciaoliv, M.; Ibanez, H.; Navarromartinez, P.C.; Perezmarcos, M.; Perezruzafa, I.; Quispebecerra, J.I.; et al. Long-Term Dynamic in Nutrients, Chlorophyll a, and Water Quality Parameters in a Coastal Lagoon During a Process of Eutrophication for Decades, a Sudden Break and a Relatively Rapid Recovery. *Front. Mar. Sci.* **2019**, *6*, 26. [CrossRef]

22. Tsikas, D. Analysis of nitrite and nitrate in biological fluids by assays based on the Griess reaction: Appraisal of the Griess reaction in the l-arginine/nitric oxide area of research. *J. Chromatogr. B Biomed. Sci. Appl.* **2007**, *851*, 51–70. [CrossRef] [PubMed]

23. Nóbrega, L.N.N.; de O. Magalhaes, L.; Fonseca, A. A urethane-acrylate microflow-analyser with an integrated cadmium column. *Microchem. J.* **2013**, *110*, 553–557.

24. Shi, Y.; Liu, H.L.; Zhu, X.Q.; Zhu, J.M.; Zuo, Y.F.; Yang, Y.; Jiang, F.H.; Sun, C.J.; Zhao, W.H. Optofluidic differential colorimetry for rapid nitrite determination. *Lab Chip* **2018**, *18*, 2994–3002. [CrossRef] [PubMed]

25. Ellis, P.S.; Shabani, A.M.H.; Gentle, B.S.; McKelvie, I.D. Field measurement of nitrate in marine and estuarine waters with a flow analysis system utilizing on-line zinc reduction. *Talanta* **2011**, *84*, 98–103. [CrossRef]

26. Cardoso, T.M.; Garcia, P.T.; Coltro, W.K. Colorimetric determination of nitrite in clinical, food and environmental samples using microfluidic devices stamped in paper platforms. *Anal. Methods* **2015**, *7*, 7311–7317. [CrossRef]

27. Liu, R.; Ishimatsu, R.; Yahiro, M.; Adachi, C.; Nakano, K.; Imato, T. Photometric flow injection determination of phosphate on a PDMS microchip using an optical detection system assembled with an organic light emitting diode and an organic photodiode. *Talanta* **2015**, *132*, 96–105. [CrossRef]

28. Nightingale, A.M.; Hassan, S.; Evans, G.W. Nitrate measurement in droplet flow: Gas-mediated crosstalk and correction. *Lab Chip* **2018**, *18*, 1903–1913. [CrossRef]

29. Sieben, V.J.; Floquet, C.F.; Ogilvie, I.R.; Mowlem, M.C.; Morgan, H. Microfluidic colourimetric chemical analysis system: Application to nitrite detection. *Anal. Methods* **2010**, *2*, 484–491. [CrossRef]

30. Cogan, D.; Fay, C.; Boyle, D.; Osborne, C.; Kent, N.; Cleary, J.; Diamond, D. Development of a low cost microfluidic sensor for the direct determination of nitrate using chromotropic acid in natural waters. *Anal. Methods* **2015**, *7*, 5396–5405. [CrossRef]

31. Yang, K.; Peretzsoroka, H.; Liu, Y.; Lin, F. Novel developments in mobile sensing based on the integration of microfluidic devices and smartphones. *Lab Chip* **2016**, *16*, 943–958. [CrossRef]

32. Beaton, A.D.; Sieben, V.J.; Floquet, C.F.; Waugh, E.M.; Ogilvie, I.R.; Mowlem, M.C.; Morgan, H. An automated microfluidic colourimetric sensor applied in situ to determine nitrite concentration. *Sens. Actuators B* **2011**, *156*, 1009–1014. [CrossRef]

33. Beaton, A.D.; Cardwell, C.L.; Thomas, R.S.; Sieben, V.J.; Legiret, F.E.; Waugh, E.M.; Statham, P.J.; Mowlem, M.C.; Morgan, H. Lab-on-Chip Measurement of Nitrate and Nitrite for In Situ Analysis of Natural Waters. *Environ. Sci. Technol.* **2012**, *46*, 9548–9556. [CrossRef] [PubMed]

34. Nightingale, A.M.; Hassan, S.; Warren, B.M.; Warren, K.; Makris, K.; Evans, G.W. A Droplet Microfluidic-Based Sensor for Simultaneous in Situ Monitoring of Nitrate and Nitrite in Natural Waters. *Environ. Sci. Technol.* **2019**, *53*, 9677–9685. [CrossRef] [PubMed]

35. Gary, H. Chemiluminescent and bioluminescent methods in analytical chemistry. A review. *Analyst* **1983**, *108*, 1274–1296.

36. Yaqoob, M.; Biot, B.F.; Nabi, A.; Worsfold, P.J. Determination of nitrate and nitrite in freshwaters using flow injection with luminol chemiluminescence de tection. *Luminescence* **2012**, *27*, 419–425. [CrossRef] [PubMed]

37. Dombrowsky, L.J.; Pratt, E.J. Fluorimetric method for determining nanogram quantities of nitrite ion. *Anal. Chem.* **1972**, *44*, 2268–2272.

38. Guo, Y.X.; Zhang, Q.F.; Shang, X. Spectrofluorimetric determination of trace nitrite with o-phenylenediamine enhanced by hydroxypropyl-β-cyclodextrin. *Spectrochim. Acta Part A* **2013**, *101*, 107–111. [CrossRef]

39. Lu, L.; Chen, C.; Zhao, D.; Yang, F.; Yang, X. A simple and sensitive assay for the determination of nitrite using folic acid as the fluorescent probe. *Anal. Methods* **2015**, *7*, 1543–1548. [CrossRef]

40. Wang, X.; Adams, E.; Schepdael, A.V. A fast and sensitive method for the determination of nitrite in human plasma by capillary electrophoresis with fluorescence detection. *Talanta* **2012**, *97*, 142–144. [CrossRef]

41. Armijo, F.; Goya, M.C.; Reina, M.; Canales, M.J.; Arévalo, M.C.; Aguirre, M.J. Electrocatalytic oxidation of nitrite to nitrate mediated by Fe (III) poly-3- aminophenyl porphyrin grown on five different electrode surfaces. *J. Mol. Catal. A Chem.* **2007**, *268*, 148–154. [CrossRef]

42. Wang, X.; Li, H.; Wu, M.; Ge, S.L.; Zhu, Y.; Wang, Q.J.; He, P.G.; Fang, Y.Z. Simultaneous Electrochemical Determination of Sulphite and Nitrite by a Gold Nanoparticle/Graphene-Chitosan Modified Electrode. *Chin. J. Anal. Chem.* **2013**, *41*, 1232–1237. [CrossRef]

43. Jońca, J.; Fernández, V.L.; Thouron, D. Phosphate determination in seawater: Toward an autonomous electrochemical method. *Talanta* **2011**, *87*, 161–167. [CrossRef] [PubMed]

44. Dahlloef, I.; Svensson, O.; Torstensson, C. Optimising the determination of nitrate and phosphate in sea water with ion chromatography using experimental design. *J. Chromatogr. A* **1997**, *771*, 163–168. [CrossRef]

45. Carrozzino, S.; Righini, F. Ion chromatographic determination of nutrients in sea water. *J. Chromatogr. A* **1995**, *706*, 277–280. [CrossRef]

46. He, L.; Zhang, K.; Wang, C.; Luo, X.; Zhang, S. Effective indirect enrichment and determination of nitrite ion in water and biological samples using ionic liquid dispersive liquid–liquid microextraction combined with high-performance liquid chromatography. *J. Chromatogr. A* **2011**, *1218*, 3595–3600. [CrossRef]

47. Campos, C.D.; Silva, J.A. Applications of autonomous microfluidic systems in environmental monitoring. *RSC Adv.* **2013**, *3*, 18216–18227. [CrossRef]

48. Wang, S.; Lin, K.; Chen, N.; Yuan, D.; Ma, J. Automated determination of nitrate plus nitrite in aqueous samples with flow injection analysis using vanadium (III) chloride as reductant. *Talanta* **2016**, *146*, 744–748. [CrossRef]

49. Khongpet, W.; Pencharee, S.; Puangpila, C.; Hartwell, S.K.; Lapanantnoppakhun, S.; Jakmunee, J. Exploiting an automated microfluidic hydrodynamic sequential injection system for determination of phosphate. *Talanta* **2017**, *177*, 77–85. [CrossRef]

50. Duffy, G.; Regan, F. Recent developments in sensing methods for eutrophying nutrients with a focus on automation for environmental applications. *Analyst* **2017**, *142*, 4355–4372. [CrossRef]

51. Legiret, F.E.; Sieben, V.J.; Woodward, E.M.S.; Bey, S.K.A.K.; Mowlem, M.C.; Connelly, D.P.; Achterberg, E.P. A high performance microfluidic analyser for phosphate measurements in marine waters using the vanadomolybdate method. *Talanta* **2013**, *116*, 382–387. [CrossRef]

52. Duffy, G.; Maguire, I.; Heery, B.; Nwankire, C.; Ducree, J.; Regan, F. PhosphaSense: A fully integrated, portable lab-on-a-disc device for phosphate determination in water. *Sens. Actuators B* **2017**, *246*, 1085–1091. [CrossRef]

53. Clinton-Bailey, G.; Beaton, A.; Nightingale, A.; Owsianka, D.; Slavik, G. A lab-on-chip analyser for in situ measurement of soluble reactive phosphate: Improved phosphate blue assay and application to fluvial monitoring. *Environ. Sci. Technol.* **2017**, *51*, 9989–9995. [CrossRef] [PubMed]

54. Ruzicka, J.; Hansen, E.H. Flow injection analyses Part I. A new concept of fast continuous flow analysis. *Anal. Chim. Acta* **1975**, *78*, 145–157.

55. Kennedy, J.F.; Weetman, D.A. A modified spectrophotometric method for the simple rapid determination of phosphate. *Anal. Chim. Acta* **1971**, *55*, 418–449. [CrossRef]

56. Grand, M.M.; Clinton-Bailey, G.S.; Beaton, A.D.; Schaap, A.M.; Johengen, T.H.; Tamburri, M.N.; Connelly, D.P.; Mowlem, M.C.; Achterberg, E.P. A Lab-On-Chip Phosphate Analyser for Long-term In Situ Monitoring at Fixed Observatories: Optimization and Performance Evaluation in Estuarine and Oligotrophic Coastal Waters. *Front. Mar. Sci.* **2017**, *4*, 255. [CrossRef]

57. Cao, X.; Zhang, S.W.; Chu, D.Z.; Wu, N.; Ma, H.K.; Liu, Y. A design of spectrophotometric microfluidic chip sensor for analyzing silicate in seawater. *IOP Conf. Ser.: Earth Environ. Sci.* **2017**, *82*, 012080. [CrossRef]

58. Feely, R.A.; Sabine, C.L.; Lee, K.; Berelson, W.; Klepas, J.; Fabry, V.J.; Millero, F.J. Impact of anthropogenic CO_2 on the $CaCO_3$ system in the oceans. *Science* **2004**, *305*, 362–366. [CrossRef]

59. Doney, S.C.; Fabry, V.J.; Feely, R.A.; Kleypas, J.A. Ocean acidification: The other CO_2 problem. *Annu. Rev. Mar. Sci.* **2009**, *1*, 169–192. [CrossRef]

60. Mostofa, K.M.; Liu, C.; Zhai, W. Reviews and Syntheses: Ocean acidification and its potential impacts, on marine ecosystems. *Biogeosciences* **2016**, *13*, 1767–1786. [CrossRef]

61. Orr, J.C.; Fabry, V.J.; Aumont, O. Anthropogenic ocean acidification over the twenty-first century and its impact on calcifying organisms. *Nature* **2005**, *437*, 681–686. [CrossRef]

62. Botelho, R.G.; Tornisielo, V.L.; de Olinda, R.A.; Maranho, L.A.; Machado-Neto, L.; Botelho, R.G.; Tornisielo, V.L.; De Olinda, R.A. Acute toxicity of sugarcane vinasse to aquatic organisms before and after pH adjustment. *Toxicol. Environ. Chem.* **2012**, *94*, 2035–2045. [CrossRef]

63. Wencel, D.; Abel, T.; McDonagh, C. Optical Chemical pH Sensors. *Anal. Chem.* **2014**, *86*, 15–29. [CrossRef] [PubMed]

64. DeGrandpre, M.D.; Spaulding, R.S.; Newton, J.O.; Jaqueth, E.J.; Hamblock, S.E.; Umansky, A.A.; Harris, K.E. Considerations for the measurement of spectrophotometric pH for ocean acidification and other studies. *Limnol. Oceanogr. Methods* **2014**, *12*, 830–839.

65. Zhang, H.; Byrne, R.H. Spectrophotometric pH measurements of surface seawater at in-situ conditions: Absorbance and protonation behavior of thymol blue. *Mar. Chem.* **1996**, *52*, 17–25. [CrossRef]

66. Rérolle, V.M.C.; Floquet, C.F.A.; Harris, A.J.K.; Mowlem, M.C.; Bellerby, R.R.G.J.; Achterberg, E.P. Development of a colorimetric microfluidic pH sensor for autonomous seawater measurements. *Anal. Chim. Acta* **2013**, *786*, 124–131. [CrossRef] [PubMed]

67. Robert-Baldo, G.L.; Morris, M.J.; Byrne, R.H. Spectrophotometric determination of seawater pH using phenol red. *Anal. Chim.* **1985**, *57*, 2564–2567. [CrossRef]

68. Byrne, R.H.; Breland, J.A. High precision multiwavelength pH determinations in seawater using cresol red. *Deep-Sea Res. Part A* **1989**, *36*, 803–810. [CrossRef]

69. Clayton, T.D.; Byrne, R.H. Spectrophotometric seawater pH measurements: Total hydrogen ion concentration scale calibration of m-cresol purple and at-sea results. *Deep-Sea Res. Part A* **1993**, *40*, 2115–2129. [CrossRef]

70. Florea, L.; Fay, C.; Lahiff, E.; Phelan, T.; O'Connor, N.E.; Corcoran, B.; Diamond, D.; Benito-Lopez, F. Dynamic pH mapping in microfluidic devices by integrating adaptive coatings based on polyaniline with colorimetric imaging techniques. *Lab Chip* **2013**, *13*, 1079–1085. [CrossRef]

71. Sumriddetchkajorn, S.; Chaitavon, K.; Intaravanne, Y. Mobile device-based self-referencing colorimeter for monitoring chlorine concentration in water. *Sens. Actuators B* **2013**, *182*, 592–597. [CrossRef]

72. Wang, Y.; Liu, X.; Chen, P. Smartphone spectrometer for colorimetric biosensing. *Analyst* **2016**, *141*, 3233–3238. [CrossRef] [PubMed]

73. Ma, J.; Yuan, D.; Lin, K.; Feng, S.; Zhou, T.; Li, Q. Applications of flow techniques in seawater analysis: A review. Trends Environ. *Anal. Chem.* **2016**, *10*, 1–10.

74. Tran-Ngoc, K.T.; Dinh, N.T.; Nguyen, T.H.; Roem, A.J.; Schrama, J.W.; Verreth, J.A. Interaction between dissolved oxygen concentration and diet composition on growth, digestibility and intestinal health of Niletilapia (Oreochromis niloticus). *Aquaculture* **2016**, *462*, 101–108. [CrossRef]

75. Wu, S.; Wu, S.; Yi, Z.; Zeng, F.; Wu, W.; Qiao, Y.; Tian, Y. Hydrogel-based fluorescent dual pH and oxygen sensors loaded in 96-well plates for high-throughput cell metabolism studies. *Sensors* **2018**, *18*, 564. [CrossRef] [PubMed]

76. Carpenter, J.H. The accuracy of the winkler method for dissolved oxygen analysis. *Limnol. Oceanogr.* **1965**, *10*, 135–140. [CrossRef]

77. Wilkin, R.T.; Mcneil, M.S.; Adair, C.J.; Wilson, J.T. Field measurement of dissolved oxygen: A comparison of methods. *Groundw. Monit. Remediat.* **2001**, *21*, 124–132. [CrossRef]

78. Trivellin, N.; Barbisan, D.; Badocco, D.; Pastore, P.; Meneghesso, G.; Meneghini, M.; Zanoni, E.; Belgioioso, G.; Cenedese, A. Study and development of a fluorescence based sensor system for monitoring oxygen in wine production: The WOW project. *Sensors* **2018**, *18*, 1130. [CrossRef]

79. Mahoney, E.J.; Hsu, H.L.; Du, F.; Xiong, B.; Selvaganapathy, P.R.; Fang, Q. Optofluidic Dissolved Oxygen Sensing With Sensitivity Enhancement Through Multiple Reflections. *IEEE Sens. J.* **2019**, *19*, 10452–10460. [CrossRef]

80. Sakai, T.; Piao, S.; Teshima, N.; Kuroishi, T.; Grudpan, K. Flow injection system with in-line Winkler' sprocedure using 16-way valve and spectrofluorimetric determination of dissolved oxygen in environmental waters. *Talanta* **2004**, *63*, 893–898. [CrossRef]

81. Zheng, G.; Xu, Z. A new design of high-precision dissolved oxygen sensor. *Transducer Microsyst. Technol.* **2012**, *2*, 112–114.

82. Norvell, W.A.; Welc, R.M. Biological Monitoring of Heavy Metal Pollution: Land and Air. *J. Environ. Qual.* **1984**, *13*, 173–174. [CrossRef]

83. Rainbow, P.S. Trace metal accumulation in marine invertebrates: Marine biology or marine chemistry. *J. Mar. Biol. Assoc. UK* **1997**, *77*, 195–210. [CrossRef]

84. Fonseca, A.; Raimundo, I.M.; Rohwedder, J.J.R.; Lima, R.S.; Araújo, M. A microfluidic device with integrated fluorimetric detection for flow injection analysis. *Anal. Bioanal. Chem.* **2010**, *396*, 715–723. [CrossRef] [PubMed]

85. Karthikeyan, K.; Sujatha, L. Fluorometric Sensor for Mercury Ion Detection in a Fluidic MEMS Device. *IEEE Sens. J.* **2018**, *18*, 5225–5231. [CrossRef]

86. Lin, M.; Hu, X.; Pan, D.; Han, H.T. Determination of iron in seawater: From the laboratory to in situ measurements. *Talanta* **2018**, *188*, 135–144. [CrossRef]

87. Heller, A.; Ronitz, O.; Barkleit, A. Complexation of Europium(III) with the Zwitterionic Form of Amino Acids Studied with Ultraviolet-Visible and Time-Resolved Laser-Induced Fluorescence Spectroscopy. *Caryologia* **2010**, *69*, 285–294. [CrossRef]

88. Milani, A.; Statham, P.J.; Mowlem, M.C. Development and application of a microfluidic in-situ analyser for dissolved Fe and Mn in natural waters. *Talanta* **2015**, *136*, 15–22. [CrossRef]

89. Provin, C.; Fukuba, T.; Okamura, K.; Fujii, T. An Integrated Microfluidic System for Manganese Anomaly Detection Based on Chemiluminescence: Description and Practical Use to Discover Hydrothermal Plumes Near the Okinawa Trough. *IEEE J. Ocean. Eng.* **2013**, *38*, 178–185. [CrossRef]

90. Lace, A.; Ryan, D.; Bowkett, M.; Cleary, J. Chromium Monitoring in Water by Colorimetry Using Optimised 1,5Diphenylcarbazide Method. *Int. J. Environ. Res. Public Health* **2019**, *16*, 1803. [CrossRef]

91. Lace, A.; Ryan, D.; Bowkett, M.; Cleary, J. Arsenic detection in water using microfluidic detection systems based on the leucomalachite green method. *Anal. Methods* **2019**, *11*, 5431–5438. [CrossRef]

92. Nguyen, H.L.; Naresh, L.; Jean-Frederic, A.; Ghasemi, R.; Lefevre, J.P.; Ha-Thi, M.H.; Mongin, C.; Leray, I. Water-soluble aluminium fluorescent sensor based on aggregation-induced emission enhancement. *New J. Chem.* **2019**, *43*, 15302–15310. [CrossRef]

93. Zhang, H.; Faye, D.; Jean-Pierre Lefèvre, J.; Delaire, J.A.; Leray, I. Selective fluorimetric detection of cadmium in a microfluidic device. *Microchem. J.* **2013**, *106*, 167–173. [CrossRef]

94. Kim, H.N.; Ren, W.X.; Kim, J.S.; Yoon, J. Fluorescent and colorimetric sensors for detection of lead, cadmium, and mercury ions. *Chem. Soc. Rev.* **2012**, *41*, 3210–3244. [CrossRef] [PubMed]

95. Xu, X.Y.; Yan, B. Eu(III) functionalized Zr-based metal-organic framework as excellent flourescent probe for Cd2+ detection in aqueous environment. *Sens. Actuators B* **2016**, *222*, 347–353. [CrossRef]

96. Siangproh, W.; Chailapakul, O.; Songsrirote, K. Simple and fast colorimetric detection of inorganic arsenic selectively adsorbed onto ferrihydrite-coated silica gel using silver nanoplates. *Talanta* **2016**, *153*, 197–202. [CrossRef] [PubMed]

97. Sikorowska, C.; Wichrowska, B. Comparative studies on the determination of zinc in water by the colorimetric method. *Roczniki Państwowego Zakadu Higieny* **1971**, *22*, 17–24.

98. Liu, H.; Liu, T.; Li, J.; Zhang, Y.; Li, J.; Song, J.; Qu, J.; Wong, W.-Y. A simple Schiff base as dual-responsive fluorescent sensor for bioimaging recognition of Zn^{2+} and Al^{3+} in living cells. *J. Mater. Chem.* **2018**, *6*, 5435–5442. [CrossRef]

99. Jo, T.G.; Bok, K.H.; Han, J.; Lim, M.H.; Kim, C. Colorimetric detection of Fe^{3+} and Fe^{2+} and sequential fluorescent detection of Al^{3+} and pyrophosphate by an imidazole-based chemosensor in a near-perfect aqueous solution. *Dyes Pigments* **2017**, *139*, 136–147. [CrossRef]

100. Saini, A.K.; Natarajan, K.; Mobin, S.M. A new multitalented azine ligand: Elastic bending, single-crystal-to-single-crystal transformation and a fluorescence turn-on Al(III) sensor. *Chem. Commun.* **2017**, *53*, 9870–9873. [CrossRef]

101. Neupane, L.N.; Mehta, P.K.; Oh, S.; Park, S.H.; Lee, K.H. Ratiometric red-emission fluorescence detection of Al^{3+} in pure aqueous solution and live cells by a fluorescent peptidyl probe using aggregation-induced emission. *Analyst* **2018**, *143*, 5285–5294. [CrossRef]

102. Zhao, L.; Wu, T.; Lefèvre, J.; Leray, I.; Delaire, J.A. Fluorimetric lead detection in a microfluidic device. *Lab Chip* **2009**, *9*, 2818–2823. [CrossRef] [PubMed]

103. Fukuba, T.; Noguchi, T.; Okamura, K.; Fujii, T. Adenosine Triphosphate Measurement in Deep Sea Using a Microfluidic Device. *Micromachines* **2018**, *9*, 370. [CrossRef] [PubMed]

104. Yamahara, K.M.; Preston, C.M.; Birch, J.M.; Walz, K.; Marin, R.; Jensen, S.; Pargett, D.; Roman, B.; Ussler, W.; Zhang, Y.W.; et al. In situ Autonomous Acquisition and Preservation of Marine Environmental DNA Using an Autonomous Underwater Vehicle. *Front. Mar. Sci.* **2019**, *6*, 373. [CrossRef]

micromachines

MDPI

Article

Silver Nanoprism Enhanced Colorimetry for Precise Detection of Dissolved Oxygen

Yunfeng Zuo [1,2], Longfei Chen [1], Xuejia Hu [1], Fang Wang [1] and Yi Yang [1,2,*]

[1] Key Laboratory of Artificial Micro- and Nano-Structures of Ministry of Education, School of Physics and Technology, Wuhan University, Wuhan 430072, China; zuoyf@whu.edu.cn (Y.Z.); coiyhm@whu.edu.cn (L.C.); huxuejiawuda@whu.edu.cn (X.H.); wangfang2200@whu.edu.cn (F.W.)
[2] Shenzhen Research Institute of Wuhan University, Shenzhen 518057, China
*   Correspondence: yangyiys@whu.edu.cn; Tel.: +86-027-68752989-8103

Received: 18 March 2020; Accepted: 2 April 2020; Published: 4 April 2020

Abstract: Dissolved oxygen (DO) content is an essential indicator for evaluating the quality of the water body and the main parameter for water quality monitoring. The development of high-precision DO detection methods is of great significance. This paper reports an integrated optofluidic device for the high precision measurement of dissolved oxygen based on the characteristics of silver nanoprisms. Metal nanoparticles, especially silver nanoprisms, are extremely sensitive to their surroundings. In glucose and glucose oxidase systems, dissolved oxygen will be transformed into H_2O_2, which affects the oxidation and erosion process of nanoprisms, then influences the optical properties of nanoparticles. By detecting the shift in the plasma resonance peak of the silver nanoparticles, the dissolved oxygen (DO) content can be determined accurately. Great reconfigurability is one of the most significant advantages of the optofluidic device. By simply adjusting the flow rate ratio between the silver nanoprisms flow and the water sample flow, real-time continuous adjustment of the detection ranges of DO from 0 to 16 mg/L can be realized dynamically. The detection limit of this device is as low as 0.11 µM (3.52 µg/L) for DO measurement. Thus, the present optofluidic system has a wide range of potential applications in fields of biomedical analyses and water sensing.

Keywords: optofluidics; dissolved oxygen; silver nanoprisms; colorimetry

---

1. Introduction

Dissolved oxygen provides the necessary biochemical environment for the survival of aquatic organisms and is an indispensable material for aquatic life activities [1,2]. Dissolved oxygen monitoring plays an important role in aquatic ecosystem quality assessment, aquatic science experiments and aquatic resource exploration [3,4]. Besides, dissolved oxygen is also a key parameter in on-chip biochemical systems and chemical processing [5]. Generally, the three primary testing methods for the determination of dissolved oxygen in water are the iodometric method, the electrode polarography method (so-called Clark electrode method) and the fluorescence lifetime method. The iodometric method has a wide detection range and high detection accuracy [6]. However, the iodometric method has the defects of cumbersome operation, long operation time and high professional requirements for the operator. The electrode polarography method is simple and fast [7,8]. However, the presence of telluride, oil, carbonate and algae in the water sample may cause clogging or even damage to the gas permeable membrane. In the actual measurement process, it requires frequent maintenance. The fluorescence lifetime method is a simple and fast approach [9]. However, the system-building is complicated, and the price is relatively high. Therefore, an economical, efficient, accurate and simple measurement of dissolved oxygen is of great value for meeting the needs of production and scientific research. For this reason, it is of vital importance to develop an accurate detection system for dissolved oxygen monitoring.

Metal nanoparticles in different shapes and sizes have various unique properties [10], which enable them to be applied in diverse fields such as biosensors and nanomedicine [11]. Compared with other kinds of metal nanoparticles, silver nanoparticles have a stronger plasmonic interaction with light [12]. Particularly, triangular silver nanoprisms exhibit more LSPR bands due to their anisotropic morphology than spherical or quasi-spherical silver nanoparticles [13]. The extraordinary properties of silver nanoprisms make it a powerful tool for biosensing and molecule detecting [10,14,15]. Optofluidics is a new interdisciplinary research field that focuses on the amalgamation of optics and microfluidics [16]. This new field provides lots of unique advantages for enhancing the performance and simplifying the design of micro-electromechanical systems [17]. Over the past decades, this new field has been developing rapidly, and it has been applied in many areas such as biosensor [18], water purification [19], optical devices [20,21] and water sensing [22]. Nowadays, the field of soft optical materials has received wide attention [23]. Optical devices based on frontier soft optical materials exhibit high biocompatibility and special qualities [24]. It is promising and expected to broaden the horizon of optofluidics in biomedical utility [25]. Optofluidics integrates multiple disciplines, which can not only realize on-chip sample pretreatment, but also can be used for the rapid detection of various biochemical indicators quickly and accurately [26].

The combination of silver nanoparticles and optofluidic systems can bring new merits. In one way, optofluidics provides a stable condition for mental nanoparticles synthesis [27]. In the other way, silver nanoparticles will strengthen the capability of optofluidic systems for biochemical sensing [28]. Therefore, optofluidics and nanotech are perfect candidates to innovate the measurement of dissolved oxygen. Herein, we designed an integrated optofluidic chip combining nanoparticle synthesis and spectrum detection to realize the precision measurement of dissolved oxygen. Rapid synthesis of silver nanoprisms was realized in a single chip. Owing to the novel physical properties of silver nanoprisms, minute amounts of dissolved oxygen concentration changes in water can be detected. Compared to traditional methods, this optofluidic DO detector possesses advantages such as higher measurement precision (detection limit less than 3.52 µg/L), less sample and reagents consumption (level of microliter), small size, low cost, parallel processing of samples, adjustable detection range (0–16 mg/L), and easy integration. This method can overcome the deficiencies of the traditional equipment in practical applications and has an important application prospect in the field of water environment monitoring.

2. Working Principle and Design

As illustrated in Figure 1, this integrated optofluidic system based on Ag-nanoprisms etching consists of three major parts: a silver nanoprisms synthesis module, a sample processing module and an optical detection module. The synthesis module was designed to realize the rapid, on-chip synthesis of silver nanoprisms. The geometric and optical features of silver nanoprisms can be controlled by adjusting the flow rate ratio of synthetic components rapidly and continuously. Through fast mixing in the microfluidic channel and precise control of reactant flows, flash chemistry can be achieved to realize rapid synthesis of nanoparticles in a single chip [29,30]. Compared with traditional batch synthesis methods, microfluidic methods process better synthesis efficiency [31]. The second part, samples mixing and processing module, is designed to realize fast and efficient mixing of pretreated water sample and synthesized Ag nanoprisms based on special microchannel design and fluid dynamics. Before being injected into the optofluidic chip, water samples with different DO contents were treated by a H_2O_2-generating system. In the presence of glucose (GO) and glucose oxidase (GOD), dissolved oxygen in the water sample is converted to H_2O_2 with concentration directly related to DO content. The reaction process is as follows:

$$\beta \ glucose + O_2 + H_2O \xrightarrow{GOD} \beta \ glucose \ acid + H_2O_2 \tag{1}$$

where GOD is glucose oxidase that catalyzes the oxidation of β-glucose quickly and efficiently. The ratio of glucose and oxygen consumption to hydrogen peroxide product is 1:1:1. In this part, the pretreated water sample with generated hydrogen peroxide and Ag nanoprisms solution are injected into the system, and they are fully mixed and reacted in the channel. Micromixer allows reagents to be thoroughly mixed, which leads to controllable and stable reaction conditions. The Ag nanoprisms will be etched due to the presence of H_2O_2 in the pretreated water sample. The etching process is as follows:

$$2Ag + H_2O_2 \rightarrow 2Ag^+ + 2OH^- \tag{2}$$

Then, the mixture flows into the third part for optical detection. An optical detecting module combining the absorption cell is designed to capture the extinction spectrum of the eroded silver nanoparticles. The SPR (surface plasmon resonance) peak is extremely sensitive to the morphology of Ag nanoprisms. Different dissolved oxygen concentrations in the water will cause varying degrees of corrosion to the silver nanoparticles and affects the extinction spectrum, causing blue shifts of SPR peak. The detectable SPR shift signal can be used for the quantitative analysis of DO contents in water samples.

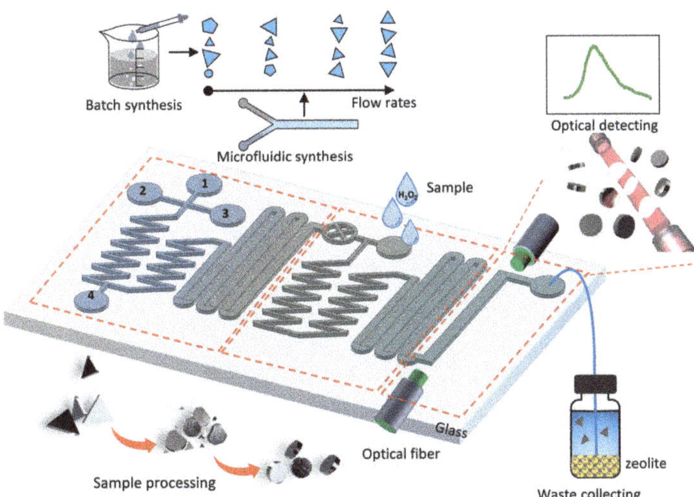

Figure 1. Schematic diagram of the optofluidic system based on Ag-nanoprisms etching for dissolved oxygen (DO) detection. This chip is consisted of three functional part: on-chip synthesis, sample processing, optical detecting. The detecting mechanism is based on the etching process of silver nanoprisms. The blue shift of the SPR peak of Ag nanoprisms can be used for the quantitative analysis of DO.

In this system, the full mixing of liquid reaction components or water samples and reagents is critical. Mixing efficiency affects not only the synthesis of nanoparticles, but also the accuracy of the system. Micromixers based on microfluidics are fundamental lab-on-a-chip components, which can realize the mixing of fluids within millisecond time scales to microsecond time scales, and have been applied in many fields [32]. Generally, micromixers can be divided into active and passive mixers. Active mixers rely on an external field, such as acoustic waves [33], to enhance the mixing and can achieve relatively high mixing efficiency [34]. Passive mixers realize fluid mixing based on structure design of the microchannel. Passive mixers are relatively simple and easy to integrate [35]. Among them, zigzag microchannel and microcylinders are usually applied to design passive micromixers based on chaotic fluid transformations [35,36]. Here, a passive micromixer based on a high-density Z-type (zigzag-type) hybrid structure was introduced, while microcylinders were designed in the Z-shape microchannel to accelerate the liquid mixing efficiency, ensuring that the reactants or samples were thoroughly mixed and reacted with each other quickly. Liquids mixing in microchannel with different designs were simulated by finite element method (FEM), as shown in Figure A1 (see Appendix A). According to the simulation, the Z-shaped channel with cylinders enables the high-speed mixing of liquids.

According to the chemical equation, the oxidization of glucose consumes O_2 and generates H_2O_2. When the concentration of glucose is insufficient to consume all the O_2 in sample water, the conversion degree of dissolved oxygen increases with glucose content. When the concentration is sufficient, the SPR peak shifts will remain the same. Under natural conditions, the concentration of DO is far less than 20 mg/L. To realize the precise detection of DO in water, a sufficient or excess amount of glucose is necessary; 20 mg O_2 needs at least 0.625 mmol glucose. The content of glucose oxidase (GOD) is relevant to the reaction efficiency and reaction time. In consideration of the response time of the whole optofluidic system and the contents of DO in actual water samples, sufficient amounts of GOD solution (5 mg/mL) and glucose solution (2 mM) were added to the water sample with the ratio of 1:1:1. To investigate the stability of sliver nanoprisms and the assay system, the effects of detection reagents, glucose, GOD, were tested. As shown in Figure A2 (see Appendix A), no detectable SPR shift was observed, showing that GO and GOD formed a stable system with Ag nanprisms.

3. Experimental Results and Discussions

3.1. Fabrication

The optofluidic device was fabricated by polydimethylsiloxane (PDMS) using soft-lithography processes. First, a 50-μm layer of SU8 photoresist (SU8-2050, Microchem, Westborough, MA, USA) was spin-coated onto a silicon wafer. After pre-baking, the master was exposed to UV light under a glass mask using a mask aligner (H94-37, SVC, Chengdu, China). Then, through the process of post-exposure bake, development and hard bake, a thick, chemically and thermally stable image was constructed on the silicon wafer as a mold. Subsequently, the mold was covered with PDMS (Sylgard 184, Dow Corning, Midland, MI, USA) and heated at 75 °C for 1 h for polymer curing. Then the cured PDMS with microchannel pattern was removed from the silicon wafer and bonded onto a glass substrate after plasma oxidation (PDC-002, Harrick Plasma, Ithaca, NY, USA). Then, the fabricated PDMS chip was stored in an oven at 75 °C for 30min to increase the bonding strength. After inserting fluidic tubing and optical fibers into the microfluidic chip, the optofluidic chip was fabricated. Reagent flow streams were injected by using micropumps (LSP01-2A, LongerPump, Baoding, China). The width of the inlet and mixing channel was 100 μm, and the width of the main channel for the reaction was 300-μm. The height of the microchannel is 50 μm. A large image was obtained by inverted microscope (Ti-E, Nikon, Tokyo, Japan) and microscope image stitching technology (NIS-Elements AR, Nikon, Tokyo, Japan) to illustrate the fabricated optofluidic device, as shown in Figure A3a (see Appendix A). Figure A3b shows the microscopes of the channel design at the positions of the inlets, zigzag-channel, microcylinders and absorption cell.

3.2. Fast Mixing

The performance of micromixers in the optofluidic systems is related to the control of Ag-nanoprisms synthesis and the reaction process between regents. To visualize the process of fast mixing in the microchannel, laser scanning confocal microscopy (A1R, Nikon, Tokyo, Japan) was used to capture the three-dimensional distribution of liquids in the microchannel. The core flow and sheath flows were dyed with Rhodamine B and Rhodamine 6G, respectively. The core flow emitted red fluorescence, and the sheath flows emitted green fluorescence. Figure 2a,b show the confocal images of liquids mixing process in the microchannel. The flow rates of the three liquid flows were 20 μL/min, respectively. The Z-shaped channel with cylinders induced the secondary flows in microchannel and increased the mixing efficiency [36]. Figure 2c,d show the intensity distribution profiles of the dotted lines in Figure 2a,b. It was obvious that the liquids were thoroughly mixed in the Z-shape channel. This microchannel design can realize the quick mixing of liquid reagents at a low flow rate, and the mixing time is about 40 ms.

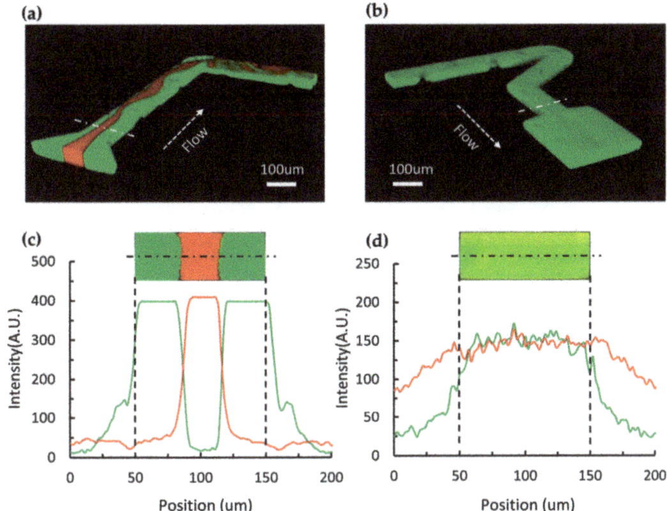

Figure 2. Mixing process of regent flows in microchannel. (**a**,**b**) Images of the three-dimensional liquids distributions capture by laser scanning confocal microscopy before and after the mixing process. The flow rates are fixed at 20 µL/min for each reagent inlet. (**c**,**d**) Cross-sectional liquids distribution corresponding to (**a**,**b**).

3.3. On-Chip Synthesis of Silver Nanoprisms

The silver nanoprisms were synthesized real-time and dynamically in the optofluidic chip according to a standard synthetic procedure at room temperature and in a neutral pH environment. [37]. Briefly, there are four fluid inlets in the synthesis module, three of which (i_1, i_2, i_3) are at the front of the efficient micro-mixer. Another inlet (i_4) was designed at the middle of the mixer. Silver nitrate ($AgNO_3$, 0.4 mmol/L), sodium citrate (4 mmol/L), and H_2O_2 (0.6 wt%) were injected into the optofluidic device through inlet i_1, i_2, i_3 with different flow rates (Qi_1 = 20 µL/min, Qi_2 = 30 µL/min, Qi_3 = 10 µL/min), respectively. These three streams were fully and quickly mixed as soon as they flowed through the micro-mixer. Then sodium borohydride ($NaBH_4$, 4 mmol/L) was injected through inlet i_4 with flow rate at 20 µL/min.

Through the synthesis module, on-chip synthesis of silver nanoprisms was realized, the geometric and optical features of silver nanoprisms can be controlled by adjusting the ratio of synthetic components rapidly and continuously. Silver nanoprisms with various diameters and SPR peaks can be synthesized through lab-on-chip systems, shown in Figure 3a. Figure 3b–e shows the TEM (Transmission Electron Microscope) micrographs of the Ag nanoprisms under different synthesis conditions. The TEM images were obtained using a JEM-2010 HT transmission electron microscope. Various shapes of silver nanoprisms were synthesized using the microfluidic method by simply changing the flow rate ratio between each regent (Qi_1:Qi_2:Qi_3:Qi_4 = 3:3:1:2, 3:3:1.5:2, 2:3:1:2), as shown in Figure 3b–d. Figure 3e shows the batch synthesis of silver nanoprisms in macroscale with the same reagents concentration as that in Figure 3d. The particle size distribution of the Ag nanoprisms synthesized by means of microfluidics and batch approach were studied, as shown in Figure A4 (see Appendix A). Owing to fast mixing in the microchannel and the precise control of reactant flows, the silver nanoprisms synthesized through the microfluidic method possess excellence reproducibility and narrow size distribution compared with batch synthesis method. The resultant Ag-nanoprisms solution was incubated for 24 h at room temperature before use, in order to eliminate the influence of borohydride, hydrogen peroxide and air bubbles. Then, the Ag nanoprisms solution was injected into the sample processing and optical detecting modules with an adjustable flow rate for DO measurement application.

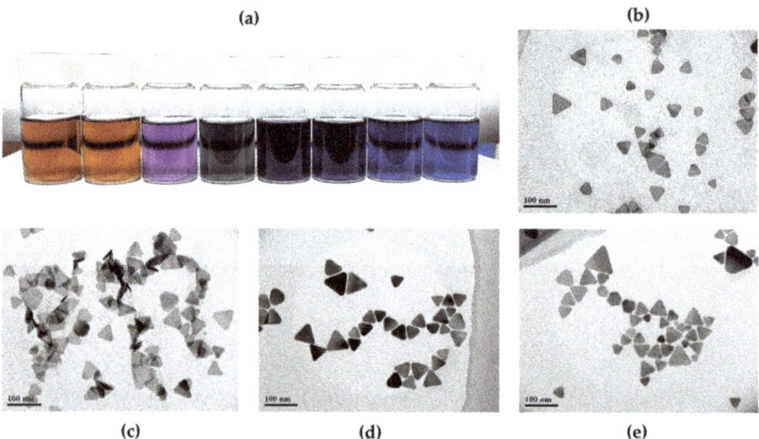

Figure 3. Synthesis of nanoprisms with different diameters. (**a**) The photograph of AgNPs synthesized through the lab-on-chip systems. TEM micrographs of the Ag nanoprisms under different synthesis conditions. Microfluidic synthesis with different flow rate ratios: $Qi_1:Qi_2:Qi_3:Qi_4$ = (**b**) 3:3:1:2, (**c**) 3:3:1.5:2, (**d**) 2:3:1:2. (**e**) Batch synthesis in macroscale.

3.4. Procedures for Dissolved Oxygen Sensing

To begin with, water samples with different dissolved oxygen concentrations were pretreated by injecting GOD solutions (5 mg/mL), glucose solution (2 mM), and oxygen-free water. The water sample with different DO contents and oxygen-free water were prepared through the aeration process by ultra-pure nitrogen and oxygen based on Henry's law. Besides, all the reagents were prepared in a nitrogen glove box by using oxygen-free water and sealed hermetically before use to minimize the impact on detection accuracy. The pretreatment process was accomplished in a Z-shape microchip. The flow rate ratio between sample water, glucose, GOD and oxygen-free water was fixed at 1:1:1:7. Then, the mixed solutions were stored into an airtight glass syringe and incubated at room temperature for 20 min. According to the chemical reaction formula, DO was transformed into H_2O_2. Then, the post-treated water samples and Ag nanoprisms solution were injected into the optofluidic chip with different flow rate ratio ($r = Q_{NPs}/Q_{WS}$) according to the concentration of DO content. Under the condition of different flow rate ratio, the total flow rate was fixed at 60 µL/min. The reaction mixture solutions were mixed and reacted along with the microchannel. H_2O_2 etches the Ag-nanprisms to silver ions. Then the solution flowed through the absorption cell. A supercontinuum whitelaser (WhiteLase SC400, Fianium, Southampton, UK) was applied to be the light source for colorimetric, and the optical signal was received and transmitted by a multimode optical fiber. Then, the SPR spectra of the mixtures were measured by a spectrograph system including a CCD camera (Newton 920, Andor, Oxford, UK) with Andor's line of Shamrock imaging spectrographs (Shamrock 303i, Andor, Oxford, UK).

Figure 4 shows the images and spectra of the Ag nanoprisms illuminated by supercontinuum whitelaser. Scattering images of silver nanoprisms before and after DO measurement (50 µM, 1.6 mg/L) are shown in Figure 4a,b. Their corresponding normalized extinction spectra captured through the optical detection module are shown in Figure 4c,d. The flow rates of post-treated water samples and Ag nanoprisms solution were fixed at Q_{WS} = 30 µL/min and Q_{NPs} = 30 µL/min, respectively (r = 1).

The spectra of the sample flow were collected 5 times, 10 s apart. By detecting the SPR peak shifts, the concentration of dissolved oxygen in the water will be determined. TEM was employed to characterize the morphological transition of Ag nanoprisms after DO sensing. The shape of Ag nanoprisms changed owing to the etching, as shown in Figure A5 (see Appendix A).

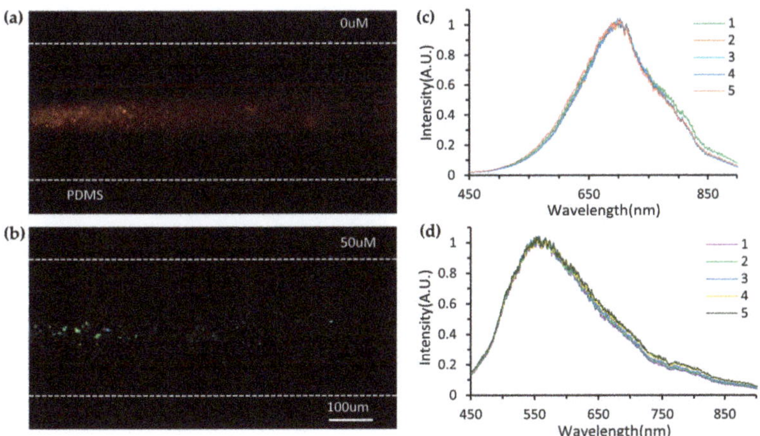

Figure 4. Images and spectra of the Ag nanoprisms illuminated by supercontinuum whitelaser. (**a**,**b**) Scattering images of silver nanoprisms before and after DO measurement (50 μM, 1.6 mg/L). And their corresponding normalized extinction spectra captured through the optical detection module (**c**,**d**). The spectra of the sample flow were collected five times, 10 s apart.

3.5. Device Performance

In glucose and glucose oxidase systems, dissolved oxygen will be transformed into H_2O_2 and further affects the oxidation and erosion process of Ag nanoprisms. The precise and sensitive detection of dissolved oxygen is practicable based on the detection of blue shift in the plasma resonance peak of the Ag nanoprisms. The shifts of the SPR peak of silver nanoprisms were measured under different dissolved oxygen concentrations, and the integration time of the spectrometer was 20 ms, as shown in Figure 5a. The blue shift of the SPR peak increased with the concentration of DO. The flow rate ratio of Ag-nanoprisms solution and the post-treatment water sample ($r = Q_{NPs}/Q_{WS}$) was fixed at 1. The corresponding relationship between the SPR peak shift ($\Delta\lambda$) and concentration (C) of DO was shown in Figure 5b. A linear relationship was formed between the SPR peak shift and DO concentration ranging from 0 to 50 μM (1.6 mg/L). The linear relationship was expressed as $\Delta\lambda = 2.3794C + 3.721$ with a correlation coefficient R^2 of 0.9889. By changing the flow rate ratio, $r = 5$, another linear calibration curve was acquired based on the SPR peak shift versus DO concentration, ranging from 0 to 250 μM (8 mg/L), as shown in Figure 5c. The linear relationship was expressed as $\Delta\lambda = 0.4554C + 9.0308$ with a correlation coefficient R^2 of 0.9944. High reconfigurability is one of the most significant advantages of optofluidic systems. By simply adjusting the ratio (r) between silver nanoprisms flow and sample flow, the continuous adjustment of the detection ranges of DO from 0 to 16 mg/L can be realized dynamically, shown in Figure 5d. The blue line shows the relationship between SPR peak shift ($\Delta\lambda$) and flow rate ratio (r) when the concentration of DO was fixed at 50 μM. The orange line shows the detection range adjustment by changing r. The insert is an enlarged graph of the relationship between detection range and r (0–1). When detecting a water sample with the same DO concentration, Ag nanoprisms will be etched to a greater degree at a low flow rate ratio, which leads to a more obvious blue shift of SPR peak. A better detection limit can be achieved due to the greater response and sensitivity of the system at a low flow rate ratio. For the case of $r = 0.5$, a relatively low detection limit can be achieved. Under the established experimental conditions, a blank water sample was measured. After 10 measurements, taking the signal-to-noise ratio of 3, the system detection limit was calculated to be $D = 3\sigma/slope = 0.11$ µM (3.52 µg/L); σ is the standard deviation of the blank sample.

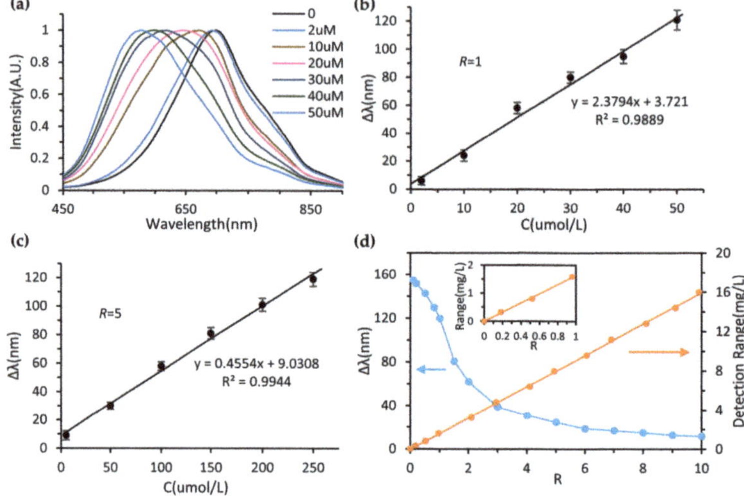

Figure 5. (**a**)Normalized SPR extinction spectra of the Ag nanoprism with different DO concentrations. Plots of SPR peak shift ($\Delta\lambda$) vs. concentrations of DO at different flow rates, (**b**) $r == 1$, (**c**) $r = 5$. (**d**) The blue line shows the relationship between SPR peak shift ($\Delta\lambda$) and flow rate ratio (r) when the concentrations of DO was fixed at 50 µM (1.6 mg/L). The orange line shows the detection range adjustment by changing r. The insert is an enlarged graph of the relationship between detection range and r (0–1).

The spectral characteristics of the silver nanoprisms under different pH values of sample water were measured, and the SPR peak shift as a function of pH was obtained. The effect of pH on the measurement results is shown in Figure 6a. Under different temperature conditions, the optofluidic device was also used to measure water with the same dissolved oxygen content. The change in the resonance peak of the silver nanoprisms was observed and recorded, as shown in Figure 6b. Under the circumstance of increasing pH value and decreasing temperatures, the SPR peak shift ($\Delta\lambda$) decreased. This was probably due to the impact on the activity of GOD under a high pH value and low-temperature condition. Besides, the detection system was not affected by pH value and temperature in a wide range, and exhibited relatively high stability.

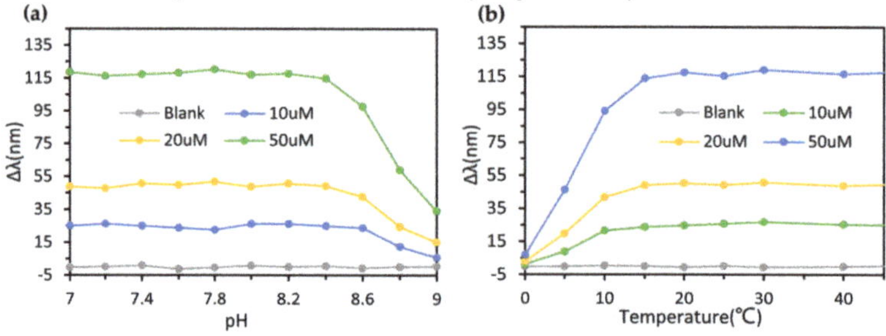

Figure 6. SPR peak shift of triangular silver nanoprisms under different conditions. (**a**) SPR peak shift of Ag nanoprisms versus pH value of sample water. (**b**) SPR peak shift of Ag nanoprisms versus incubation temperature.

Since the ratio of nanoprisms and the water sample of the optofluidic device can be changed, a wide and adjustable detection range can be achieved. Compared with other iodometric method, electrode polarography method, and fluorescence method, the optofluidic method based on silver nanoprisms possesses better performance in dissolved oxygen detection Table 1 lists the performance of different DO detection techniques.

Table 1. Comparison of optofludic method with other DO detection techniques.

Assay Methods	Sensitivity	Range	Detection Limit	Reference
Optofluidic DO detector	$7.5\ nm\cdot L\cdot mg^{-1}$	$0–16\ mg\cdot L^{-1}$	$3.52\ \mu g\cdot L^{-1}$	This work
Iodometric method	–	$0\text{-}20\ mg\cdot L^{-1}$	$0.1\ mg\cdot L^{-1}$	[6]
Electrode polarography	$5.5\ \mu A\ L\cdot mg^{-1}$	$0.2\text{-}6.5\ mg\cdot L^{-1}$	$0.02\ mg\cdot L^{-1}$	[8]
Fluorescence lifetime	$I_0/I = 117$	$0\text{-}40\ mg\cdot L^{-1}$	–	[9]

In addition, the waste solution containing silver after DO measurement was collected and treated by zeolite [38]. The wastewater treatment process is shown in Figure A6 (see Appendix A). Then the absorbed silver was recycled by the precipitation method [39]. The rate of recovery can reach up to 98.3%, and the purity of silver powder can achieve 99.8%. The recovered silver can be recycled in DO measurement. This saves reagents, reduces detection cost and avoids environmental pollution.

4. Conclusions

In conclusion, we demonstrated a high-accuracy, eco-friendly optofluidic dissolved oxygen detector based on the etching of silver nanoprisms. The characteristics of the system and its anti-interference ability were investigated. For the measurement of dissolved oxygen, the detection limit is as low as $3.52\ \mu g/L$. The device possesses repeatability and optical stability. It is suitable for the measurement of dissolved oxygen under different environmental conditions. Compared with traditional dissolved oxygen determination methods, the optofluidic detector possesses a higher measurement resolution with a lower cost. Besides, the optofluidic device has advantages such as less reagent consumption and simple operation, and can realize precise detection and high adjustability in its detection range. High precision and reproducibility in the detection of DO is beneficial for researchers in helping them discover deep phenomena, conduct aquatic analysis, and reveal new laws in water environmental science. Therefore, the optofluidic dissolved oxygen detector has a good application prospect in water quality analysis.

Under the established conditions, the accurate detection of DO has been realized. In order to illustrate the effect of more factors such as the kinetics of mixing, total flow rate, micromixer design on the performance of the optofluidic DO sensor, further systematical investigation is necessary and is under conducting in our laboratory. Besides metal nanoparticles, there are still some materials, such as frontier soft optical materials, that are worthy of exploitation and have great potential in terms of biomedical sensing. This study will be beneficial to simplifying the system and optimizing the chip structure for various sensing applications.

Author Contributions: Conceptualization, Y.Z. and Y.Y.; methodology, Y.Z.; writing—original draft preparation, Y.Z.; writing—review and editing, L.C., X.H. and F.W.; supervision, Y.Y; all authors contributed to manuscript writing. All authors have read and agreed to the published version of the manuscript.

Funding: This work was financially supported by the Foundation Research Fund of Shenzhen Science and Technology Program (no. JCYJ20170818112939064), the National Natural Science Foundation of China (no. 11774274), the Open Financial Grant from Qingdao National Laboratory for Marine Science and Technology (no. QNLM2016ORP0410), National Key R&D Program of China (2018YFC1003200). We also acknowledge the assistance with nanofabrication provided by the Center for Nanoscience and Nanotechnology at Wuhan University.

Acknowledgments: The authors would like to thank Xiaoqiang Zhu for beneficial discussions, and Yu Gao for manuscript preparation.

Conflicts of Interest: The authors declare no conflict of interest. The funders had no role in the design of the study; in the collection, analyses, or interpretation of data; in the writing of the manuscript, or in the decision to publish the results.

Appendix A

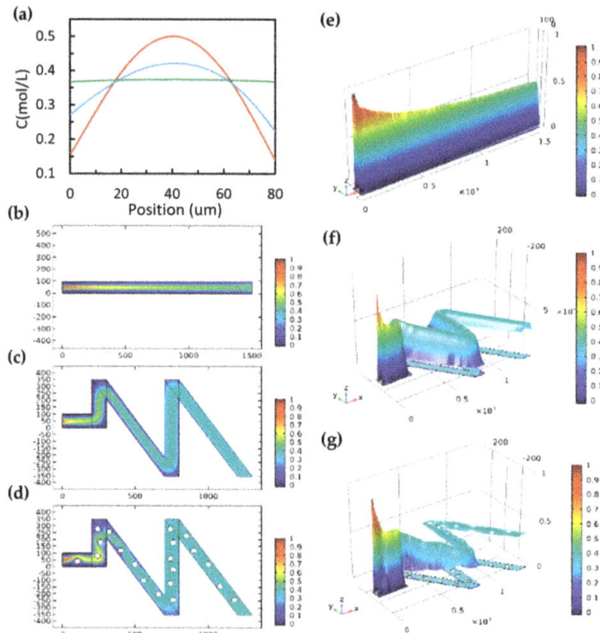

Figure A1. Simulation of liquids mixing in microchannel with different designs. (**a**) Cross-sectional concentration profiles of straight channel (red line), Z-shape channel (blue line) and Z-shape channel with blocks (green line) at x = 1000 μm. (**b**–**d**) Stable concentration distribution in main channel by simulation. (**e**–**g**) Corresponding height curves.

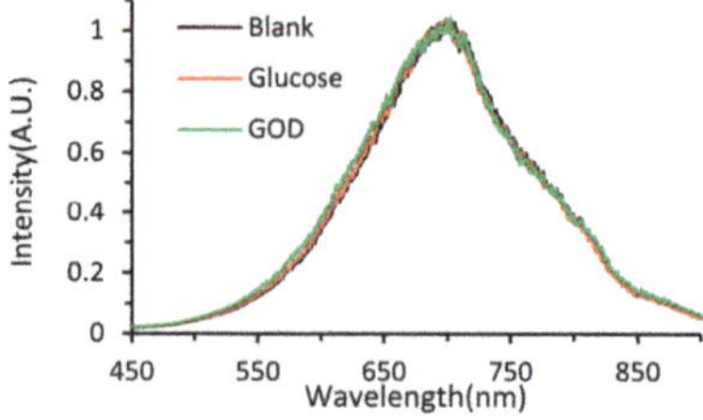

Figure A2. SPR extinction spectra of Ag nanoprisms in presence of glucose or GOD.

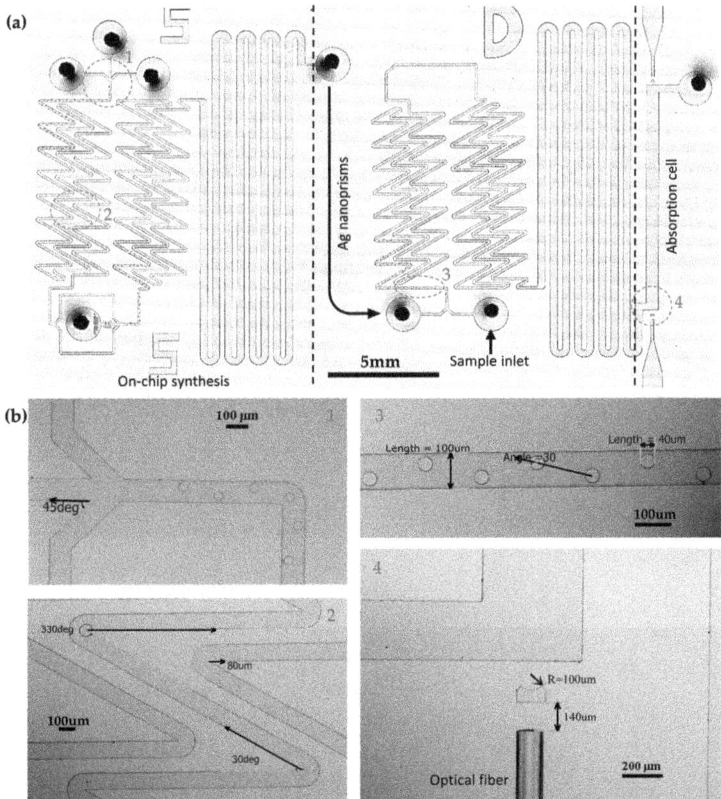

Figure A3. (a)Large Image of the fabricated chip obtained by image stitching technology and microscopes of the channel design at the positions of inlets (**b1**), zigzag-channel (**b2**), microcylinders (**b3**) and absorption cell (**b4**).

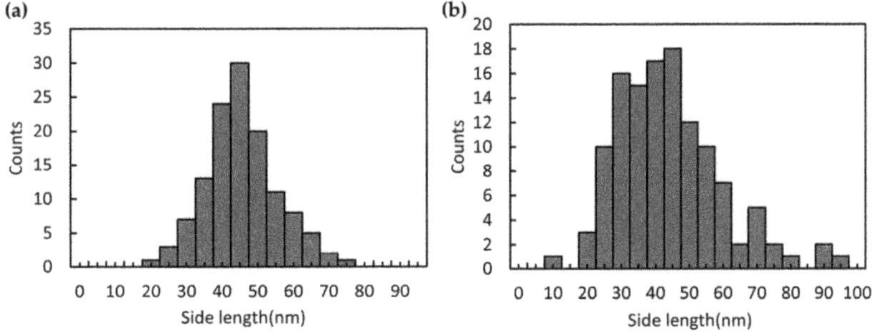

Figure A4. Particle size distribution of the Ag nanoprisms by means of microfluidics (**a**) and batch approach (**b**).

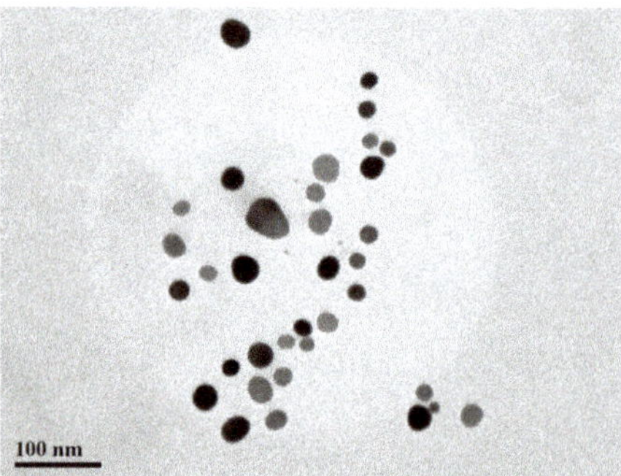

Figure A5. TEM micrographs of the Ag nanoprisms after the detection of dissolved oxygen (50 μM, 1.6 mg/L), the shape of Ag nanoprisms changed owing to the etching.

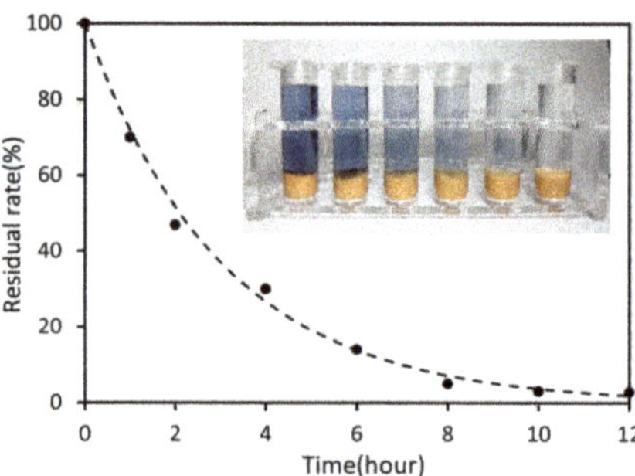

Figure A6. Absorption of the silver waste water by zeolite. Residual rate of silver nanoparticles as a function of time during the water purifying process.

References

1. Diaz, R.J.; Rosenberg, R. Spreading dead zones and consequences for marine ecosystems. *Science* **2008**, *321*, 926–929. [CrossRef] [PubMed]
2. Papkovsky, D.B.; Dmitriev, R.I. Biological detection by optical oxygen sensing. *Chem. Soc. Rev.* **2013**, *42*, 8700–8732. [CrossRef] [PubMed]
3. Stramma, L.; Schmidtko, S.; Levin, L.A.; Johnson, G.C. Ocean oxygen minima expansions and their biological impacts. *Deep Sea Res. Part I Oceanogr. Res. Pap.* **2010**, *57*, 587–595. [CrossRef]
4. Whitehead, P.G.; Wilby, R.L.; Battarbee, R.W.; Kernan, M.; Wade, A.J. A review of the potential impacts of climate change on surface water quality. *Hydrol. Sci. J.* **2009**, *54*, 101–123. [CrossRef]
5. Oomen, P.E.; Skolimowski, M.D.; Verpoorte, E. Implementing oxygen control in chip-based cell and tissue culture systems. *Lab Chip* **2016**, *16*, 3394–3414. [CrossRef]
6. Shriwastav, A.; Sudarsan, G.; Bose, P.; Tare, V. A modified Winkler's method for determination of dissolved oxygen concentration in water: Dependence of method accuracy on sample volume. *Measurement* **2017**, *106*, 190–195. [CrossRef]
7. Luo, J.; Dziubla, T.; Eitel, R. A low temperature co-fired ceramic based microfluidic Clark-type oxygen sensor for real-time oxygen sensing. *Sens. Actuators B Chem.* **2017**, *240*, 392–397. [CrossRef]
8. Damos, F.S.; Luz, R.C.; Tanaka, A.A.; Kubota, L.T. Dissolved oxygen amperometric sensor based on layer-by-layer assembly using host–guest supramolecular interactions. *Anal. Chim. Acta* **2010**, *664*, 144–150. [CrossRef]
9. Chu, C.S.; Lo, Y.L. Optical fiber dissolved oxygen sensor based on Pt (II) complex and core-shell silica nanoparticles incorporated with sol–gel matrix. *Sens. Actuators B Chem.* **2010**, *151*, 83–89. [CrossRef]

10. Zhang, Z.; Wang, H.; Chen, Z.; Wang, X.; Choo, J.; Chen, L. Plasmonic colorimetric sensors based on etching and growth of noble metal nanoparticles: Strategies and applications. *Biosens. Bioelectron.* **2018**, *114*, 52–65. [CrossRef]

11. Kumar, A.; Kim, S.; Nam, J.M. Plasmonically engineered nanoprobes for biomedical applications. *J. Am. Chem. Soc.* **2016**, *138*, 14509–14525. [CrossRef] [PubMed]

12. Millstone, J.E.; Hurst, S.J.; Metraux, G.S.; Cutler, J.I.; Mirkin, C.A. Colloidal gold and silver triangular nanoprisms. *Small* **2009**, *5*, 646–664. [CrossRef] [PubMed]

13. Pastoriza-Santos, I.; Liz-Marzán, L.M. Colloidal silver nanoplates. State of the art and future challenges. *J. Mater. Chem.* **2008**, *18*, 1724. [CrossRef]

14. Zhou, Y.; Huang, W.; He, Y. pH-Induced silver nanoprism etching-based multichannel colorimetric sensor array for ultrasensitive discrimination of thiols. *Sens. Actuators B Chem.* **2018**, *270*, 187–191. [CrossRef]

15. Xia, Y.; Ye, J.; Tan, K.; Wang, J.; Yang, G. Colorimetric visualization of glucose at the submicromole level in serum by a homogenous silver nanoprism-glucose oxidase system. *Anal. Chem.* **2013**, *85*, 6241–6247. [CrossRef] [PubMed]

16. Psaltis, D.; Quake, S.R.; Yang, C. Developing optofluidic technology through the fusion of microfluidics and optics. *Nature* **2006**, *442*, 381–386. [CrossRef]

17. Zuo, Y.F.; Zhu, X.Q.; Shi, Y.; Liang, L.; Yang, Y. Light manipulation in inhomogeneous liquid flow and its application in biochemical sensing. *Micromachines* **2018**, *9*, 163.

18. Liang, L.; Zuo, Y.F.; Wu, W.; Zhu, X.Q.; Yang, Y. Optofluidic restricted imaging, spectroscopy and counting of nanoparticles by evanescent wave using immiscible liquids. *Lab Chip* **2016**, *16*, 3007–3014. [CrossRef]

19. Wang, N.; Zhang, X.; Wang, Y.; Yu, W.; Chan, H.L. Microfluidic reactors for photocatalytic water purification. *Lab Chip* **2014**, *14*, 1074–1082. [CrossRef]

20. Zhu, X.Q.; Liang, L.; Zuo, Y.F.; Zhang, X.M.; Yang, Y. Tunable visible cloaking using liquid diffusion. *Laser Photonics Rev.* **2017**, *11*, 1700066. [CrossRef]

21. Chen, Q.M.; Jian, A.Q.; Li, Z.H.; Zhang, X.M. Optofluidic tunable lenses using laser-induced thermal gradient. *Lab Chip* **2016**, *16*, 104–111. [CrossRef] [PubMed]

22. Zhu, J.M.; Shi, Y.; Zhu, X.Q.; Yang, Y.; Jiang, F.H.; Sun, C.J.; Zhao, W.H.; Han, X.T. Optofluidic marine phosphate detection with enhanced absorption using a Fabry-Perot resonator. *Lab Chip* **2017**, *17*, 4025–4030. [CrossRef] [PubMed]

23. Huang, W.; Ling, S.; Li, C.; Omenetto, F.G.; Kaplan, D.L. Silkworm silk-based materials and devices generated using bio-nanotechnology. *Chem. Soc. Rev.* **2018**, *47*, 6486–6504. [CrossRef] [PubMed]

24. Sun, Y.-L.; Dong, W.-F.; Yang, R.-Z.; Meng, X.; Zhang, L.; Chen, Q.-D.; Sun, H.-B. Dynamically tunable protein microlenses. *Angew. Chem. Int. Ed.* **2012**, *51*, 1558–1562. [CrossRef] [PubMed]

25. Sun, Y.-L.; Sun, S.-M.; Wang, P.; Dong, W.-F.; Zhang, L.; Xu, B.-B.; Chen, Q.-D.; Tong, L.-M.; Sun, H.-B. Customization of protein single nanowires for optical biosensing. *Small* **2015**, *11*, 2869–2876. [CrossRef]

26. Li, Y.; Liu, C.; Feng, X.; Xu, Y.; Liu, B.F. Ultrafast microfluidic mixer for tracking the early folding kinetics of human telomere G-Quadruplex. *Anal. Chem.* **2014**, *86*, 4333–4339. [CrossRef]

27. Liu, Y.; Jiang, X.Y. Why microfluidics? Merits and trends in chemical synthesis. *Lab Chip* **2017**, *17*, 3960–3978. [CrossRef]

28. Ma, J.P.; Lee, S.M.Y.; Yi, C.Q.; Li, C.W. Controllable synthesis of functional nanoparticles by microfluidic platforms for biomedical applications—A review. *Lab Chip* **2017**, *17*, 209–226. [CrossRef]

29. Yoshida, J.; Nagaki, A.; Yamada, T. Flash chemistry: Fast chemical synthesis by using microreactors. *Chemistry* **2008**, *14*, 7450–7459. [CrossRef]

30. Carboni, M.; Capretto, L.; Carugo, D.; Stulz, E.; Zhang, X.L. Microfluidics-based continuous flow formation of triangular silver nanoprisms with tuneable surface plasmon resonance. *J. Mater. Chem. C* **2013**, *1*, 7540. [CrossRef]

31. Abou-Hassan, A.; Sandre, O.; Cabuil, V. Microfluidics in inorganic chemistry. *Angew. Chem.* **2010**, *49*, 6268–6286. [CrossRef] [PubMed]

32. DeMello, A.J. Control and detection of chemical reactions in microfluidic systems. *Nature* **2006**, *442*, 394–402. [CrossRef] [PubMed]

33. Ahmed, D.; Mao, X.; Juluri, B.K.; Huang, T.J. A fast microfluidic mixer based on acoustically driven sidewall-trapped microbubbles. *Microfluid. Nanofluid.* **2009**, *7*, 727–731. [CrossRef]

34. Ozcelik, A.; Ahmed, D.; Xie, Y.; Nama, N.; Qu, Z.; Nawaz, A.A.; Huang, T.J. An acoustofluidic micromixer via bubble inception and cavitation from microchannel sidewalls. *Anal. Chem.* **2014**, *86*, 5083–5088. [CrossRef] [PubMed]

35. Liu, C.; Li, Y.; Liu, B.F. Micromixers and their applications in kinetic analysis of biochemical reactions. *Talanta* **2019**, *205*, 120136. [CrossRef]

36. Amini, H.; Sollier, E.; Masaeli, M.; Xie, Y.; Ganapathysubramanian, B.; Stone, H.A.; Di Carlo, D. Engineering fluid flow using sequenced microstructures. *Nat. Commun.* **2013**, *4*, 1826. [CrossRef]

37. Zhang, Q.; Li, N.; Goebl, J.; Lu, Z.; Yin, Y. A systematic study of the synthesis of silver nanoplates: Is citrate a "magic" reagent? *J. Am. Chem. Soc.* **2011**, *133*, 18931–18939. [CrossRef]

38. Akgul, M.; Karabakan, A.; Acar, O.; Yurum, Y. Removal of silver (I) from aqueous solutions with clinoptilolite. *Microporous Mesoporous Mater.* **2006**, *94*, 99–104. [CrossRef]

39. Syed, S. Silver recovery aqueous techniques from diverse sources: Hydrometallurgy in recycling. *Waste Manag.* **2016**, *50*, 234–256. [CrossRef]

Article

Structural Stability of Optofluidic Nanostructures in Flow-Through Operation

Yazan Bdour [1,†], **Juan Gomez-Cruz** [1,2,†] **and Carlos Escobedo** [1,*]

1   Department of Chemical Engineering, Queen's University, Kingston, ON K7L 3N6, Canada;
    16yb6@queensu.ca (Y.B.); 17jmgc@queensu.ca (J.G.-C.)
2   Instituto de Ciencias Aplicadas y Tecnología (ICAT), Universidad Nacional Autónoma de México (UNAM),
    Ciudad de México 04510, Mexico
*   Correspondence: ce32@queensu.ca
†   These authors contributed equally to work.

Received: 25 February 2020; Accepted: 31 March 2020; Published: 2 April 2020

Abstract: Optofluidic sensors based on periodic arrays of subwavelength apertures that support surface plasmon resonance can be employed as both optical sensors and nanofluidic structures. In flow-through operation, the nanoapertures experience pressure differences across the substrate in which they are fabricated, which imposes the risk for structural failure. This work presents an investigation of the deflection and structural stability of nanohole array-based optofluidic sensors operating in flow-through mode. The analysis was approached using experiments, simulations via finite element method, and established theoretical models. The results depict that certain areas of the sensor deflect under pressure, with some regions suffering high mechanical stress. The offset in the deflection values between theoretical models and actual experimental values is overturned when only the effective area of the substrate, of 450 μm, is considered. Experimental, theoretical, and simulation results suggest that the periodic nanostructures can safely operate under trans-membrane pressures of up to 20 psi, which induce deflections of up to ~20 μm.

Keywords: optofluidic; sensor; surface plasmon resonance; nanohole array; mechanical properties; nanofluidic; nanoplasmonic

---

1. Introduction

The development of new point-of-care (POC) diagnostic technologies requires low-cost, fully integrated sensing platforms capable of providing quantitative results in situ. At the same time, POC diagnostic platforms have a tremendous potential that is yet to be fully exploited. Telemedicine, for instance, aims to monitor the health of patients remotely through on-site sensing using personal devices, holding a global market of ca. $20 billion USD (United States dollars) [1]. A trendy and increasingly demanded approach to in situ sensing is the use of lab-on-a-chip platforms enabled by cell phones to record, analyze, and transmit the results [2–4]. With the recent emergence of new pathogens, such as the Coronavirus and the Yaravirus, an on-site analysis will limit their health impact with a rapid sensing test, quantifying the severity of the infection, and assisting with the quarantine measures [5,6]. Periodic arrays of subwavelength structures fabricated in metal films enable surface plasmon resonance (SPR), which motivated their use as biosensors for several applications in different fields [7–12]. Ordered arrays of metallic nanoholes are optofludic structures that enable transport of both fluid and analyte via nanofluidic confinement and nanoplasmonic sensor. The plasmonic resonance signature obtained from nanohole arrays (NHAs) allows the detection of biologically relevant analytes in label-free fashion and real time. Toward the development of POC biosensing platforms, these optofluidic nanostructures are integrated into microfluidic environments in order to create fully integrated sensors compatible with portable electronics [13]. NHA-based sensors are ideal for field applications due to their small footprint and integration abilities as evidenced by recent demonstrations for the detection of bacteria, such as *Chlamydia trachomatis* [14], viruses, such as Ebola [15], cancer biomarkers [16] and uropathogenic bacteria [17]. Flow-through optofluidic structures also enable the enrichment of analytes in liquids by an electrohydrodynamic effect occurring around the NHAs when an electric potential and a pressure bias are applied to the fluid in a closed system [18]. Despite their demonstrated potential in sensing, most applications involving nanohole arrays focus on exploiting the conventional optical capacities of these nanostructures. The mechanical stability of the nanohole membranes is an overlooked aspect of their properties that are key when functioning as nanofluidic structures. In analogy to porous silicon-based membranes, where permeability increases significantly as membrane thickness decreases, the volumetric flow across nanostructured optofluidic sensors increases with the open pore fraction. However, plasmonic nanostructures with built-in thin membranes may suffer from low mechanical stability which could limit, critically, their use as optofluidic flow-through sensors [19–21]. The membrane's mechanical properties change due to the change in the structural morphology of the porous membrane as it deflects under pressure. The stability decreases by a correction factor $(1 - P)$, where P is related to the porosity of the membrane [22].

Recent studies demonstrate that through-nanoapertures fabricated in thin (~50 nm) gold-coated Si_3N_4 substrates offer additional fluidic abilities that can be used to target in-hole delivery of analytes when operated as optofluidic sensors [23,24]. However, flow-through operation results in transmembrane pressures that could potentially damage the rather brittle nanostructures. The mechanical properties of the organized nanohole arrays are not completely understood due to their sensitivity, brittleness, and nano-sized structures. Here, we present a study on structural aspects of Au-on-nitride optofluidic nanoplasmonic sensors operating in flow-through fashion at flowrates compatible with biosensing applications.

2. Materials and Methods

2.1. Fabrication of Periodic through Subwavelength Apertures

Through-nanohole arrays were fabricated using focused ion beam (FIB) milling using 100-nm-thick Si_3N_4 free-standing membranes (Norcada, Edmonton, AB, Canada) coated with a thermally evaporated 100 nm layer of gold via a 5-nm chromium adhesion layer. Milling was achieved using a gallium ion beam set at 40 keV with a beam current of ~30 pA, with a typical beam spot size of 10 nm, and the dwell time of the beam at one pixel was set to 20 µs. Two arrays of through-nanohole arrays with an area of 20 µm by 20 µm, diameter of ca. 230 nm, and pitch of 560 nm were fabricated.

2.2. Fabrication of Microfluidic Chips

The microfluidic chip was fabricated using a replica molding technique as described in detail elsewhere [25]. The general steps of the fabrication procedure are briefly described next. A mask with the microfluidic pattern was generated using SolidWorks CAD software (Dassault Systems Solidworks Corp., Waltham, MA, USA). The design included one inlet and one outlet of 1.5 mm, and a 5-mm-wide channel with 100 µm in height. A master was fabricated by spin-coating SU-8 100 photoresist (MicroChem Corp., Newton, MA, USA) on a clean three-inch silicon wafer (Silicon Quest International Inc., Santa Clara, CA, USA). The coated wafer was then prebaked for one minute at 65 °C and for 10 min at 95 °C. The mask with the channel pattern was then placed over the coated wafer and exposed to ultraviolet (UV) light for 90 s. Next, the exposed wafer was hard-baked at 65 °C for 1 min and at 95 °C for 10 min. The master was subsequently developed using a SU-8 developer (MicroChem Corp., Newton, MA, USA). A 12:1 mixture of Sylgard 184 elastomer to curing agent (Dow Corning, Midland, MI, USA) was mixed, degassed in a vacuum, and poured onto the master. After baking at 85 °C for 20 min, the replica was removed from the mold. Inlets and outlets were provided 1-mm punched holes for fluidic access. Microfluidic connections were achieved using polyether ether ketone

(PEEK) tubing (Upchurch Scientific, Oak Harbor, WA, USA). A schematic representation of the set-up is shown in Figure 1.

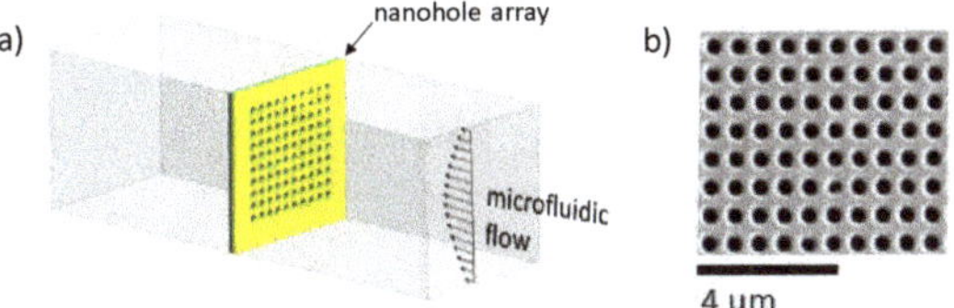

Figure 1. (**a**) SEM image of fabricated periodic subwavelength apertures via focused ion beam (FIB). The nanostructures were 230 nm in diameter and 560 nm in pitch. (**b**) Schematic representation of a nanohole array in a microfluidic chip in flow-through operation.

2.3. Optofluidic Structure Deflection Analysis

Finite element analysis (FEA) was used as a means to know the order of magnitude of the deflection and the mechanical stress that the optofluidic sensor may experience in flow-through operation. COMSOL Multiphysics (COMSOL, Stockholm, Sweden) was used to simulate a simplified model of the optofluidic sensor under a prescribed unidirectional and orthogonal pressure on one of the faces of the suspended membrane. The simulations were used firstly to estimate the order of magnitude of applied pressures that would result on the deflection of the substrate containing the optofluidic structures. This first model involved a stationary elastic model with default Lagrange–quadratic element type. The finite element analysis solves for the displacement field at a specific point on the membrane for every input force. For the linear model, the system is governed by three tensor partial differential equations: $\nabla \cdot \sigma + F_v = 0$, $\varepsilon = \frac{1}{2}\left[(\nabla u)^T + \nabla u + (\nabla u)^T \nabla u\right]$, and $C = C(E, v)$, where σ is the Cauchy stress tensor, F_v is the body force per unit, u is the displacement vector, ε is the infinitesimal strain tensor, C is the fourth-order stiffness tensor, E is the Young's modulus, and v is the Poisson's ratio. A second static, nonlinear stress–strain model was used to compare the experimental data and to validate the deflection values obtained for the prescribed pressure range. The nonlinear stress–strain behavior was achieved by using a power-law nonlinear elastic material model, accounting for geometric nonlinearities, which is governed by Ludwik's law, $\tau = \tau_0 + k\gamma^{1/n}$, where τ is the shear stress, γ is the shear strain, and n is an integer [26,27]. A user-controlled mesh with Lagrange–quadratic element type was used for this nonlinear model, to guarantee an acceptable mesh size along the thickness of the modeled substrate. The finite element analysis solves for the displacement field at a specific point on the membrane for every input force. In both models, linear and nonlinear, the parameters of Si_3N_4 were mainly used, as the values for the mechanical properties for this material supersede those of the metal components in the sensor, namely, a Young's modulus of 250×10^9 Pa, a density of 3.1×10^3 kg/m^3, and a Poisson's ratio of 0.23. The surrounding surfaces around the membrane that correspond to the areas that define the thickness of the substrate were set as fixed boundaries. The transmembrane pressures were varied from 1 to 20 psi, as this range corresponds to flow rates on the order of nL/min, which are commonly used in biosensing applications. The deflection of the substrate and the stress (von Mises criterion) were recorded.

In addition to finite element method (FEM)-based models, analytical models on the mechanical behavior of perforated membranes published in the literature were also used to estimate the deflection of the optofluidic sensors in this study, as detailed in the Section 3 [22].

3. Results and Discussion

Figure 2 shows a schematic representation of the experimental setup used to measure the deflection of the membranes. Figure 3 shows the computer-aided design (CAD) models used to study the deflection of the optofluidic sensors via COMSOL Multiphysics software. Figure 3a shows the simplified model with a single nanoaperture at the center, used in the linear elastic material simulations. The model accounts for a 100-nm-thick membrane with a square surface with a side length of 500 μm, and a circular opening of 10 μm for surface coverage equivalency of the effective surface of the nanoapertures. Figure 3b shows an image of the CAD model used for the nonlinear simulations, a square 100-nm-thick membrane with side length of 500 μm and a 20 μm × 20 μm array of 230-nm-diameter holes with pitch-to-diameter ratio of 2. In both cases, linear and nonlinear models, the mesh curvature factor was 0.6, the maximum element scaling factor was 1.9, the resolution of narrow regions was 0.3, and the optimize quality feature was set to on. The linear model had a maximum element size at all boundaries of 30×10^{-9}. The resulting mesh had ~210×10^3 domain elements with ~40×10^3 boundary elements and ~1.4×10^3 edge elements. The nonlinear model had ~1.4×10^4 domain elements, ~900×10^3 boundary elements, and ~6×10^3 edge elements. The models were solved for pressures applied to the bottom surface of the substrate, for 1 psi, and then using the sweep parameter feature for a pressure range of 2–20 psi with 2-psi pressure increments.

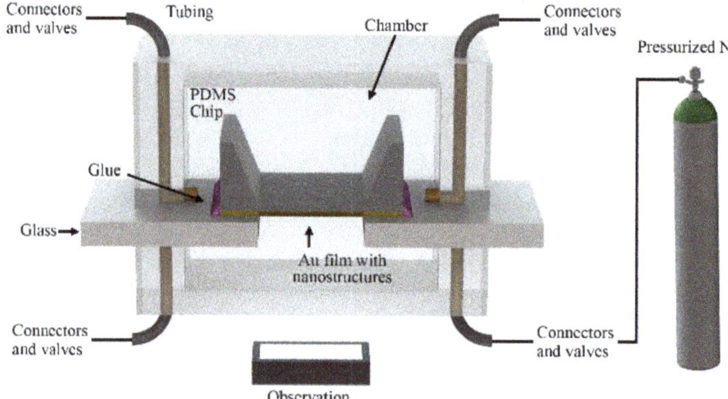

Figure 2. Schematic representation of the experimental set-up.

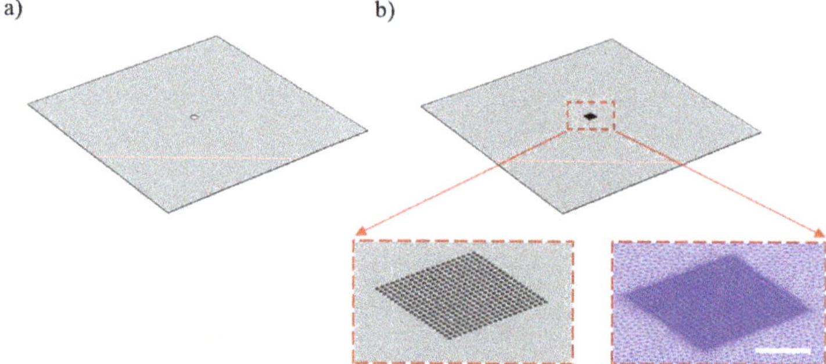

Figure 3. Computer-aided design (CAD) models used for the finite element method (FEM)-based simulations. (**a**) CAD model used for linear elastic simulations. (**b**) CAD model used for the nonlinear elastic simulations. A detail of the nanoapertures in the CAD model and the corresponding mesh are shown as insets. Scale bar represents 10 μm.

Figure 4 shows images of selected values for the deflection and stress distribution of the model of the membrane under an applied pressure of 20 psi. Figure 4a,b show the displacement in the z-direction for the linear and nonlinear models, respectively. The results are presented as non-deformed, with vectors representing the direction and magnitude of the deflection. The pattern of deflection observed from the simulations, as expected, is quasi-circular, with increasing magnitude toward the center of the free-standing membrane. The maximum deflection values, for the linear and nonlinear simulations at an applied pressure of 20 psi, were 24.08 and 19.39 μm, respectively. Maxima were always obtained at the apex of the deformed membrane. Figure 4c,d show the von Mises stress distribution for an applied pressure of 20 psi. The maximum stress found in the simulations was on the order of 1×10^8 to 10×10^8 Pa, which suggests that the substrate which is housing the nanoapertures could adequately withstand the deformations resulting from the applied pressure. The simulation results were used to define a range of pressure that could be used experimentally, avoiding failure of the membrane.

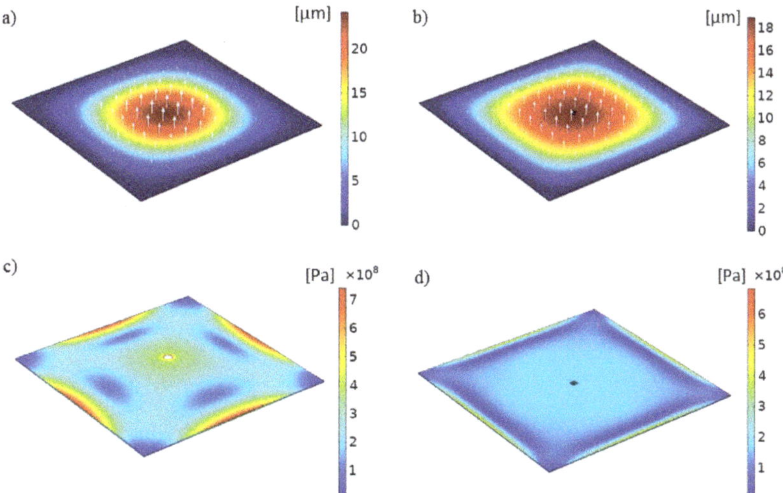

Figure 4. Simulation results of linear and nonlinear models. Membrane deflection under an applied pressure of 20 psi for (**a**) the linear model and (**b**) the nonlinear model. The apertures are shown in the insets within yellow dashed boxes in both cases. Stress distribution (von Mises yield criterion) under an applied pressure of 20 psi for (**c**) the linear model and (**d**) the nonlinear model.

Figure 5 shows a bright-field microscopy image of the Au-on-nitride membrane before and after the application of a pressure of 10 psi. The substrate included two rectangular periodic arrays of nanoapertures, indicated with yellow dashed lines. The boundaries of the Si_3N_4 membrane are indicated by red dashed lines. The focal plane in both images is the same, which indicates the deflection of the substrate under the applied pressure.

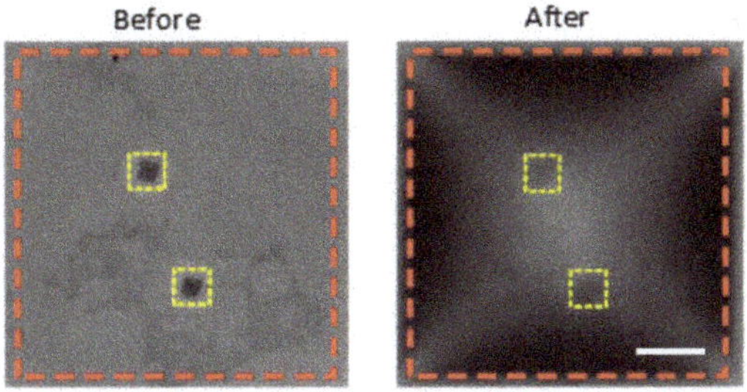

Figure 5. Membrane deflection before and after the application of a pressure of 20 psi. Scale bar represents 100 μm.

In order to measure the deflection experimentally, the elevation difference at the apex of the membrane was used as reference, and the in-focus *z*-positions were recorded. The applied pressure on the surface of the substrate was monitored and regulated to achieve a constant value throughout the measurement of the deflection. Fringe patterns can be observed in the deflected membrane case, which correspond to the reflected light, confirming a level gradient along the surface of the substrate, and a maximum translation at the apex. The *z*-positioning precision of the inverted microscopy system used in this study was 0.2 μm, which allowed measuring deflections with micrometer precision, at 2-μm intervals.

Figure 6 shows experimental and simulations results for applied pressures of 1–20 psi. The trend from the linear simulation model was linear, as expected, with corresponding minimum and maximum deflections of 1.209 and 24.08 μm. In contrast, the deflection results from the nonlinear model decreased with the applied pressure, with minimum and maximum values of 2.584 and 19.39 μm. The same trend was found for experimental values, with the magnitude of the maximum deflection at the apex decreasing with the applied pressure. This can be explained by considering the physical restriction along the frame of the free-standing membrane and due to the mechanical properties of the material. The figure also shows the results from three analytical models that were used to obtain theoretical values, i.e., the Rijn et al. [22], Ugural [28], and Kovacs et al. [29] models, as well as an adjusted Kovacs model fit with the experimental values. These models are similar to each other, whereby they all consider the perforation in a membrane as an error factor affecting the Young's modulus of the membrane. The deflection of a membrane is given by Equation (1) [30].

$$w_{max} = k_0 L \sqrt[3]{\frac{P_0 L}{E_{eff} h}},$$ (1)

where w is the z-axis displacement, L and h are the size and the thickness of the membrane, and P_0 is the applied pressure. The constant k_0 is equal to 0.318, 0.325, and 0.319 within Rijn's, Ugural's, and Kovacs' models, respectively [30]. E_{eff} is the effective Young's modulus, calculated as $E_{eff} = (1 - P)E_{closed}$, where E_{closed} is the Young's modulus of unperforated membrane, and P is the correction factor. P is dependent on the perforation and is defined as the fraction of the open areas over the total area of the membrane. As the models are similar, there is negligible difference between the deflection values obtained using the three different models [30]. In the case of the optofluidic sensor, the deformable section of the membrane is smaller than the 500 μm by 500 μm of the free-standing substrate, as observed in Figure 5. The theoretical models do not consider the frame around the deformable area. Therefore, there is an offset between the deflection values obtained using the models and those obtained experimentally, as shown in Figure 6.

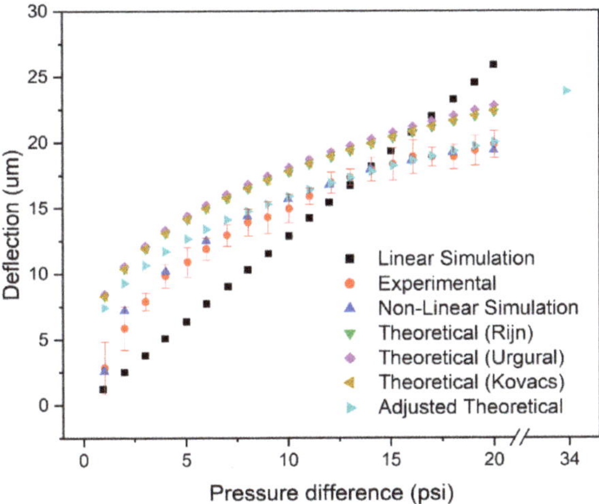

Figure 6. Experimental, theoretical, and simulation results of the maximum membrane deflection (apex). Error bars indicate standard deviation ($n = 5$).

The experimental results have a similar trend compared to the theoretical models. Over the non-linear region (< 7 psi), the experimental values are on average $\sim$ 32% below the theoretical maximum, and $\sim$ 12% below the theoretical maximum within the linear region (> 7 *psi*). The slopes for the experimental and theoretical (Kovacs) values were 0.4831 and 0.4826 µm/psi, respectively within the linear region, with R^2 (coefficient of determination (COD)) values of 0.937 and 0.993, respectively. The slopes indicate that the models do not quantify the actual deflection of the membrane. However, they accurately represent the trend of the membrane's deflection. As such, the unperforated area around the nanohole arrays is influential on the mechanical stability of the membrane. With the assumption that some length of area around the unperforated area does not deflect, then the deflection of the membrane can be rewritten as follows:

$$w_{max,} = k_0 L_{eff} \sqrt[3]{\frac{P_0\, L_{eff}}{E_{eff}\, h}},$$

(2)

where L_{eff}, is the effective length of the membrane based on the experimental values, calculated as $L_{eff} = \sqrt{Area\ of\ holes / P_{eff}}$. P_{eff} is the effective correction factor based on the experimental values, where it is assumed that some length around the unperforated area does not deflect. E_{eff} is adjusted to the experimental values and calculated as $E_{eff} = (1 - P_{eff}) E_{closed}$. The model found a range of P_{eff} values based on each experimental deflection point from 2.138×10^{-3} to 5.09×10^{-4}, corresponding to effective membrane lengths of 225 µm to 460 µm, respectively. Figure 6 illustrates that the model is incapable of fitting all the experimental values with one value of P_{eff}. The initial deflection value of the experimental values has an L_{eff} of 225 µm, where the L_{eff} non-linearly increases until it plateaus to a constant value of 460 µm within the linear region of the experimental values. The effective length paints a clear image of the membrane's behavior under pressure. Initially, at low pressures, only the center area of the membrane deflects, while the majority of the membrane is not affected by the applied pressures. As the applied pressure increases, the deflected area grows until it reaches a maximum constant value (460 µm). Even at the maximum value of effective lengths, some outer areas of the membrane do not deflect, reassuring the limitations of deflection model. The experiment was not designed to bring the substrate to mechanical failure; however, the pressure value for the breaking point can be extrapolated from the theoretical model based on the material's properties. The inflection point of the membrane is not at the edges of the membrane but limited to the effective length of the membrane (i.e., L_{eff}). Based on the Rijn et al. and Timoshenko et al. models, the maximum pressure applied can be found based on the total stress of the material as shown in Equation (3) [22,31].

$$\sigma_{total} = \sigma_{tensile} + \sigma_{bend} = \frac{0.297}{1-v}\left(1 + \frac{1.439}{0.358}\right)\sqrt[3]{\frac{P_0^2\, L_{eff}^2\, E_{eff}}{(1-v^2)\, h^2}}, \tag{3}$$

where σ_{total} is the total stress of the membrane, and $\sigma_{tensile}$ and σ_{bend} are the tensile stress due to stretching and the maximum bending stress near the middle of the membrane's deflection edges, respectively. The model is valid when the substrate is under a substantial load that results in large deflections (i.e., $w_{max}/h \gg 1$). Considering that the reported ultimate stress, $\sigma_{ultimate}$, is on the order of 10^9 Pa, and the intrinsic tensile stress is 10^8 Pa for a silicon nitride membrane, then the internal stresses can be neglected since they are an order of magnitude lower than the total stress [22]. For a nonductile inorganic material, the $\sigma_{ultimate}$ is equivalent to its yield stress. Taking 2.5 GPa as σ_{total}, based on the mechanical properties of the material, a pressure of 33.91 psi and deflection of 23.87 μm are obtained, corresponding to the maximum possible values at the verge of mechanical failure [32]. This theoretical maximum deflection value at the verge of failure, which adequately follows the trend of the adjusted theoretical curve, is shown in Figure 6.

4. Conclusions

This work presented an investigation of the deflection and structural stability of optofluidic nanohole array-based sensors operating in flow-through mode. The study was approached using experiments, theoretical models, and FEA via computer simulations through FEM. Linear and nonlinear material models were simulated using COMSOL Multiphysics software. The simplified linear model had an expected discrepancy with experimental values, but these were useful to obtain an estimation of the order of magnitude of transmembrane pressures that would allow studying the deflection of the substrate when used in flow-through operation, while avoiding mechanical failure. The discrepancies were up to ~20%. In contrast, the nonlinear model, accounting for a complete nanohole array, accurately described the deflection values obtained experimentally. The stresses corresponding to these deflections can be used to predict maximum operation values that could prevent failure of the optofluidic nanostructures. Three analytical models were used to analyze the deformation of the sensor. The models depicted the behavior of the deflected substrate under pressure but did not intrinsically fit the experimental results since only a fraction of the surface deflects due to the attachment of the free-standing substrate to the silicon frame. Even when the entire 500-μm membrane is under pressure, only a reduced square area, ranging from 225 μm to a maximum of 460 μm per side, deflects. Once adjusted, the theoretical model better fit the experimental deflection values. Based on the models, the fracture point was extrapolated from the maximum yield stress of silicon nitride membranes. As the membranes are composed of nonductile, inorganic material, their yield stress is equivalent their ultimate stress, which resulted with a maximum possible deflection of 23.9 μm, with the applied pressure of 33.9 psi. Although the optofluidic structures are limited by their fragile mechanical stability in flow-through operation, these results show that they are capable of withstanding transmembrane pressures compatible with sensing applications, where the analyte is required to be brought into the apertures. Simulations that could predict the deflection of the structures would greatly benefit the design needs of flow-through optofluidic platforms for specific applications in the context of biosensing.

Author Contributions: Conceptualization, C.E.; methodology, Y.B., J.G.-C., C.E.; software, Y.B., J.G.-C., C.E.; validation, C.E.; formal analysis, Y.B., J.G.-C., C.E.; investigation, Y.B., J.G.-C., C.E.; resources, C.E.; data curation, C.E.; writing—original draft preparation, Y.B., J.G.-C., C.E.; writing—review and editing, C.E.; supervision, C.E.; project administration, C.E.; funding acquisition, C.E. All authors have read and agreed to the published version of the manuscript.

Funding: C.E. acknowledges funding from the Natural Sciences and Engineering Research Council of Canada (NSERC) RGPIN-2010-5138, and Canada Foundation for Innovation John R. Evans Leaders' Fund Program

(No. 319670). Y.B. acknowledges funding from Materials for Advanced Photonics and Sensing (MAPS) funded through NSERC CREATE and Ontario Graduate Scholarship (OGS). J.G.-C. acknowledges the national graduate scholarship provided by the National Council for Science and Technology of Mexico (CONACYT).

Conflicts of Interest: The authors declare no conflicts of interest.

References

1. Marcoux, R.M.; Vogenberg, F.R. Telehealth: Applications From a Legal and Regulatory Perspective. *P T Peer-Rev. J. Formul. Manag.* **2016**, *41*, 567–570.
2. Preechaburana, P.; Gonzalez, M.C.; Suska, A.; Filippini, D. Surface Plasmon Resonance Chemical Sensing on Cell Phones. *Angew. Chem. Int. Ed.* **2012**, *51*, 11585–11588. [CrossRef]
3. Comina, G.; Suska, A.; Filippini, D. Autonomous chemical sensing interface for universal cell phone readout. *Angew. Chem. Int. Ed.* **2015**, *54*, 8708–8712. [CrossRef]
4. Srinivasan, B.; Tung, S. Development and applications of portable biosensors. *J. Lab. Autom.* **2015**, *20*, 365–389. [CrossRef]
5. Li, G.; De Clercq, E. *Therapeutic options for the 2019 novel coronavirus (2019-nCoV)*; Nature Publishing Group: New York, NY, USA, 2020.
6. Boratto, P.V.; Oliveira, G.P.; Machado, T.B.; Andrade, A.C.S.; Baudoin, J.-P.; Klose, T.; Schulz, F.; Azza, S.; Decloquement, P.; Chabriere, E.; et al. A mysterious 80 nm amoeba virus with a near-complete" ORFan genome" challenges the classification of DNA viruses. *bioRxiv* **2020**. [CrossRef]
7. Brolo, A.G.; Gordon, R.; Leathem, B.; Kavanagh, K.L. Surface plasmon sensor based on the enhanced light transmission through arrays of nanoholes in gold films. *Langmuir* **2004**, *20*, 4813–4815. [CrossRef]
8. Ebbesen, T.W.; Lezec, H.J.; Ghaemi, H.F.; Thio, T.; Wolff, P.A. Extraordinary optical transmission through sub-wavelength hole arrays. *Nature* **1998**, *391*, 667–669. [CrossRef]
9. Craighead, H. Future lab-on-a-chip technologies for interrogating individual molecules. In *Nanoscience And Technology: A Collection of Reviews from Nature Journals*; World Scientific: Singapore, Singapore, 2010; pp. 330–336.
10. Ahn, C.H.; Choi, J.-W.; Beaucage, G.; Nevin, J.H.; Lee, J.-B.; Puntambekar, A.; Lee, J.Y. Disposable smart lab on a chip for point-of-care clinical diagnostics. *Proc. IEEE* **2004**, *92*, 154–173. [CrossRef]
11. Herrmann, M.; Veres, T.; Tabrizian, M. Enzymatically-generated fluorescent detection in micro-channels with internal magnetic mixing for the development of parallel microfluidic ELISA. *Lab Chip* **2006**, *6*, 555–560. [CrossRef]
12. Chin, C.D.; Linder, V.; Sia, S.K. Lab-on-a-chip devices for global health: Past studies and future opportunities. *Lab Chip* **2007**, *7*, 41–57. [CrossRef]
13. Gomez-Cruz, J.; Nair, S.; Ascanio, G.; Escobedo, C. Flow-through nanohole array based sensor implemented on analogue smartphone components. In Proceedings of the SPIE Nanoscience + Engineering, San Diego, CA, USA, 25 August 2017.
14. Soler, M.; Belushkin, A.; Cavallini, A.; Kebbi-Beghdadi, C.; Greub, G.; Altug, H. Multiplexed nanoplasmonic biosensor for one-step simultaneous detection of Chlamydia trachomatis and Neisseria gonorrhoeae in urine. *Biosens. Bioelectron.* **2017**, *94*, 560–567. [CrossRef] [PubMed]
15. Yanik, A.A.; Huang, M.; Kamohara, O.; Artar, A.; Geisbert, T.W.; Connor, J.H.; Altug, H. An Optofluidic Nanoplasmonic Biosensor for Direct Detection of Live Viruses from Biological Media. *Nano Lett.* **2010**, *10*, 4962–4969. [CrossRef]
16. Escobedo, C.; Chou, Y.W.; Rahman, M.; Duan, X.B.; Gordon, R.; Sinton, D.; Brolo, A.G.; Ferreira, J. Quantification of ovarian cancer markers with integrated microfluidic concentration gradient and imaging nanohole surface plasmon resonance. *Analyst* **2013**, *138*, 1450–1458. [CrossRef]
17. Gomez-Cruz, J.; Nair, S.; Manjarrez-Hernandez, A.; Gavilanes-Parra, S.; Ascanio, G.; Escobedo, C. Cost-effective flow-through nanohole array-based biosensing platform for the label-free detection of uropathogenic E. coli in real time. *Biosens. Bioelectron.* **2018**, *106*, 105–110. [CrossRef]
18. Escobedo, C.; Brolo, A.G.; Gordon, R.; Sinton, D. Optofluidic concentration: Plasmonic nanostructure as concentrator and sensor. *Nano Lett.* **2012**, *12*, 1592–1596. [CrossRef] [PubMed]
19. Cruz, S.; Hönig-d'Orville, A.; Müller, J. Fabrication and optimization of porous silicon substrates for diffusion membrane applications. *J. Electrochem. Soc.* **2005**, *152*, C418–C424. [CrossRef]

20. Sinton, D.; Gordon, R.; Brolo, A.G. Nanohole arrays in metal films as optofluidic elements: Progress and potential. *Microfluid. Nanofluidics* **2008**, *4*, 107–116. [CrossRef]
21. Wu, H.; Liu, G.; Wang, J. Atomistic and continuum simulation on extension behaviour of single crystal with nano-holes. *Model. Simul. Mater. Sci. Eng.* **2004**, *12*, 225. [CrossRef]
22. Van Rijn, C.; van der Wekken, M.; Nijdam, W.; Elwenspoek, M. Deflection and maximum load of microfiltration membrane sieves made with silicon micromachining. *J. Microelectromech. Syst.* **1997**, *6*, 48–54. [CrossRef]
23. Escobedo, C.; Brolo, A.G.; Gordon, R.; Sinton, D. Nanoplasmonics as nanofluidics: Transport and sensing in flowthrough nanohole arrays. In Proceedings of the Microfluidics, BioMEMS, and Medical Microsystems IX, San Francisco, CA, USA, 14 February 2011.
24. Escobedo, C. On-chip nanohole array based sensing: A review. *Lab Chip* **2013**, *13*, 2445–2463. [CrossRef]
25. Duffy, D.C.; McDonald, J.C.; Schueller, O.J.A.; Whitesides, G.M. Rapid prototyping of microfluidic systems in poly(dimethylsiloxane). *Anal. Chem.* **1998**, *70*, 4974–4984. [CrossRef] [PubMed]
26. Luecke, W.E.; Wiederhorn, S.M. A new model for tensile creep of silicon nitride. *J. Am. Ceram. Soc.* **1999**, *82*, 2769–2778. [CrossRef]
27. Wunderlich, W. *Ceramic Materials*; InTech: Rijeka, Croatia, 2010.
28. Ugural, A.C. *Stresses in Beams, Plates, and Shells*; CRC press: Boca Raton, FL, USA, 2009.
29. Kovács, Á.; Vízváry, Z.; Kovács, A.; Mescheder, U. Large deflection analysis of perforated silicon nitride membranes. In Proceedings of the 6th ESMC, Budapest, Hungary, 28 August–1 September 2006.
30. Kovacs, A.; Kovács, Á.; Pogány, M.; Mescheder, U. Mechanical investigation of perforated and porous membranes for micro-and nanofilter applications. *Sens. Actuators B Chem.* **2007**, *127*, 120–125. [CrossRef]
31. Timoshenko, S.P.; Woinowsky-Krieger, S. *Theory of Plates and Shells*; McGraw-hill: New York, NY, USA, 1959.
32. Boé, A.; Coulombier, M.; Safi, A.; Pardoen, T.; Raskin, J.-P. On-chip Testing Laboratory for Nanomechanical Characterization of Thin Films. In Proceedings of the 2009 SEM Annual Conference and Exposition on Experimental and Applied Mechanics, Albuquerque, NM, USA, 1–4 June 2009.

Article

3D Hydrodynamic Focusing in Microscale Optofluidic Channels Formed with a Single Sacrificial Layer

Erik S. Hamilton [1,*], Vahid Ganjalizadeh [2], Joel G. Wright [1], Holger Schmidt [2] and Aaron R. Hawkins [1]

[1]   Electrical and Computer Engineering, Brigham Young University, Provo, UT 84602, USA;
     joel.g.wright@studentbody.byu.edu (J.G.W.); ahawkins@byu.edu (A.R.H.)
[2]   Electrical and Computer Engineering, University of California, Santa Cruz, Santa Cruz, CA 95064, USA;
     vganjali@ucsc.edu (V.G.); hschmidt@soe.ucsc.edu (H.S.)
*   Correspondence: erikshamilton@byu.edu

Received: 11 February 2020; Accepted: 26 March 2020; Published: 27 March 2020

Abstract: Optofluidic devices are capable of detecting single molecules, but greater sensitivity and specificity is desired through hydrodynamic focusing (HDF). Three-dimensional (3D) hydrodynamic focusing was implemented in 10-µm scale microchannel cross-sections made with a single sacrificial layer. HDF is achieved using buffer fluid to sheath the sample fluid, requiring four fluid ports to operate by pressure driven flow. A low-pressure chamber, or pit, formed by etching into a substrate, enables volumetric flow ratio-induced focusing at a low flow velocity. The single layer design simplifies surface micromachining and improves device yield by 1.56 times over previous work. The focusing design was integrated with optical waveguides and used in order to analyze fluorescent signals from beads in fluid flow. The implementation of the focusing scheme was found to narrow the distribution of bead velocity and fluorescent signal, giving rise to 33% more consistent signal. Reservoir effects were observed at low operational vacuum pressures and a balance between optofluidic signal variance and intensity was achieved. The implementation of the design in optofluidic sensors will enable higher detection sensitivity and sample specificity.

Keywords: 3D hydrodynamic focusing; optofluidic; lab-on-a-chip; biosensor; microscale channel; microfluidic; liquid-core waveguide; single layer; reservoir effect

---

1. Introduction

The optofluidic detection of particles in liquid waveguide channels has recently grown in significance [1–7]. The intersecting of solid-core waveguides with liquid-core waveguides enables light-matter interaction, such as fluorescence generation for single molecule detection. This is especially interesting for identification of particles in liquid, such as disease pathogens [8–14]. The small scale of the optofluidic channels is critical for single molecule detection in flow.

The nature of the microfluidic channels found on optofluidic sensor platforms sets some of the ultimate sensitivity and accuracy limits. The small cross-sectional area of the channels typically means operation in the laminar flow regime, which results in a parabolic velocity distribution. Fluid that is in the center of the channel flows faster than fluid at the walls. This means fluorescent particles, being distributed evenly through a channel, would spend different amounts of time in perpendicular excitation light beams as they flow past excitation points introducing variation in excitation times and fluorescent signal generation. Additionally, the excitation light exhibits an optical mode intensity profile that results in a distribution of generated fluorescence intensities. Additionally, the collection efficiency of fluorescence in a liquid-core waveguide is dependent on particle position [15,16]. Particles that are near the center of the channel experience higher collection efficiency. Moreover, velocity-based identification schemes suffer from the particle velocity distribution as variation limits the sensitivity and specificity capabilities of these methods. Taken together, all of these possible variations in signal intensity and particle velocity distribution introduce uncertainty in particle detection.

Hydrodynamic focusing (HDF) promises to enhance the detection capabilities in optofluidic channels by controlling the position of target particles in the fluid channel. This results in narrower velocity and excitation intensity distributions, which means less variation in the fluorescent signal intensity distribution [17]. Three-dimensional hydrodynamic focusing (3DHDF) aims to limit the position of target particles both horizontally and vertically. This ensures they pass through the excitation beam, flow at a more uniform velocity, and experience optimal collection efficiencies resulting in less fluorescent signal intensity variation. We have predicted a signal enhancement of between three and five times with 3DHDF, but the complexity of implementation should be considered.

HDF is commonly achieved using buffer fluid or sheath flows to surround and squeeze a sample fluid stream. The engineering challenge comes in designing and fabricating the intersection for these flows. Two-dimensional hydrodynamic focusing is relatively straightforward, even at the 10-µm scale. It simply requires flowing buffer fluid from the left and right sides of the sample stream to achieve horizontal, or in-plane, focusing, as in Figure 1a. The buffer fluid occupies the edges of the microfluidic channel volume, limiting the locations that sample particles (flowing in from a central channel) can occupy. Focusing the sample stream from the top and bottom, or vertically, is much more difficult to achieve, because the fluid must be directed from out-of-plane. This can lead to complex designs, difficult fabrication processes, and complicated operation. However, the combination of horizontal and vertical focusing resulting in a sample stream surrounded on all sides, or sheathed, by buffer fluid, as in Figure 1b, enables enhanced optofluidic performance and applications [18–22]. Our goal with this work was to develop a simple buffer fluid induced 3D focusing scheme at the 10-µm scale while using surface micromachining. Ultimately, we desired a design only requiring a single sacrificial layer of material to avoid complexity in fabrication and maximize device yield. We also wanted to keep device operation as simple as possible.

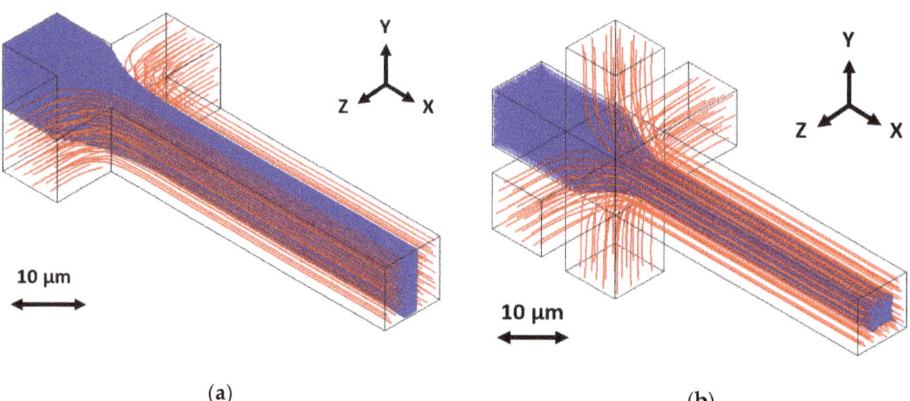

Figure 1. Hydrodynamic focusing occurs when sample fluid (drawn in solid blue) is squeezed and sheathed from (**a**) two dimensions (horizontal focusing) or (**b**) three dimensions (horizontal and vertical focusing) by a buffer fluid (drawn in striped red), controlling the position of sample fluid in the fluid channel.

A wide variety of interesting designs have been developed to perform 3DHDF while using buffer fluid at the microscale. One existing method literally injects sample fluid into a buffer fluid stream using a nanoneedle, or micronozzle [23]. The micronozzle cantilever is formed with oxide over silicon before the silicon is etched and the nanoneedle released. This fragile structure extends into the 100 µm scale microfluidic channel, where buffer fluid occupies the volume and performs the sheathing function for 3DHDF. Not only is the fabrication complex, but the structure is fragile and it might be impossible to shrink to the 10-μm scale as the oxide annealing process that forms the nozzle may lead to a closed nozzle. Alternatively, tilted lithography has been used to create fluid manifolds for directing buffer fluid around sample fluid for sheathing [24]. The fabrication process requires a rotatable, tiltable jig holder for the microfluidic device during multiple photolithography exposures. The design is similar to the micronozzle, in that it includes a sample injection nozzle. Again, the structures are on the 100 μm scale. They both require three or four fluid reservoirs to operate, depending on layout. Other designs requiring complex fabrication processes or operation include a femtosecond laser exposure design and a PDMS stack that requires five to seven layers to make and six fluid reservoirs to operate [25,26].

Simpler, "single layer" designs make use of clever fluid dynamics properties to induce 3DHDF with buffer fluid through secondary flow production. This takes the form of Dean Vortex generation in fast moving fluids. The vortices induce flow orthogonal to the main flow direction, drawing out and shaping the sample stream. The microscale fluid channel structures can be simpler because the fluid dynamics do some of the work, reducing complexity in fabrication and operation.

The curve design, sometimes called "microfluidic drifting", requires four or five fluid reservoirs to operate [27]. Fluid moves through a 90-degree curve in a microchannel. Sample fluid is located on the inside of the curve and buffer fluid on the outside. As the fast-moving fluids travel around the bend, secondary flow occurs, drawing the sample fluid out horizontally. Additional buffer fluid envelopes the vertically focused sample stream from the sides, performing horizontal focusing and completing the 3DHDF.

The contraction-expansion array design (CEA) works in a similar manner, inducing secondary flow in the form of Dean Vortex generation [28]. Rather than a curve or bend, the CEA consists of a chain of narrow and wide fluid channel sections that cause the fluid to contract and expand. The sample fluid is located on the inside of the narrow contraction regions. When it reaches the expansion region, the sample fluid is horizontally drawn out before being contracted again. The pressure change between the regions induces the secondary flow and the effect compounds. Unlike the Curve design, the CEA completes 3DHDF with no additional buffer fluid for horizontal focusing. In this case, three fluid reservoirs are required for operation.

The last of the single-layer, vortex generator designs discussed here is the microstructure stream-sculpting design [29]. More commonly known as micropillars, the structures are located within the volume of the microfluidic channel in the path of the already horizontally focused sample stream. When the sample stream comes in contact with the micropillars, its path is redirected around the structures and secondary flow occurs. The effect is much like that of the CEA where the contraction and expansion caused by the row of pillars has a compounding effect, thereby sculpting the stream from a vertical stripe to a horizontal ellipse sheathed by buffer fluid. This design requires three to four fluid reservoirs, as the sample stream must be horizontally focused prior to its contact with the microstructures.

Note that these designs are more typical at the 100-μm scale or larger. When shrunk to the 10-μm scale, the fluidic resistance of the small cross-section fluid channels dramatically increases and the flow velocities that are required to induce the secondary flow focusing effect become impractical due to the massive microfluidic channel backpressures [17]. Thus, low fluid velocity is required and the existing single-layer designs become impossible to implement. This means that we cannot rely on secondary flow to do the work of focusing. We must rely on a sheathing or "volumetric flow ratio" design that can be operated at low velocity and in small cross-section channels.

We present a 3DHDF design requiring just one layer of sacrificial material. The low-pressure chamber design, or pit, requires four fluid ports to operate and it promises improved device yield over multi-layer designs. Moreover, it operates by pressure driven volumetric flow ratio, enabling low velocity focusing in channels with dimensions around 10 μm. We will present computer simulations, fabrication processes, and optical characterization of the focusing completed in microscale channels. Fluorescent beads in a flowing aqueous solution are used to characterize the design and demonstrate the enhanced detection capabilities of the system.

2. Materials and Methods

2.1. Design Concept

We have previously reported 3DHDF at the 10-µm scale while using a volumetric flow ratio technique [30]. However, complexity in the fabrication and operation remained. The design required three consecutive layers of sacrificial SU-8 photoresist that formed the fluid channel volume. The dimension steps from layer to layer introduced crevices in the silicon dioxide overlayers that were used to form the walls of the channel leading to structurally weak devices. Moreover, the design required six fluid reservoirs to operate, taking up large amounts of chip real-estate and making operation challenging. A two-layer design was also implemented, improving the yield and decreasing operation complexity by using four fluid reservoirs [31]. However, the multi-layer structure remained less than optimal and resulted in low device yield.

Quantifying fabrication complexity is challenging, as it includes processing time and difficulty. For example, multi-layer photoresist designs are extremely difficult in that subsequent layers must be perfectly processed on the first try, whereas single-layer designs allow for layer stripping and repatterning. Additionally, each layer attempt requires between three and six hours. Moreover, multi-layer designs prove to be less robust, resulting in low device yields. Every parallel processed device can become simultaneously damaged near the end of the multi-week fabrication process, requiring complete device remanufacturing. Moving to a single-layer design greatly simplifies the fluid volume formation process.

The schematics presented in Figure 2a,b show the fluid junction region of the more easily fabricated, more robust design introduced by this paper. It is formed with one layer of sacrificial photoresist and requires four fluid reservoirs to operate by pressure-driven volumetric flow ratio. The key fluid features that cause the fluid focusing are a pit and trench pair. These are etched into the substrate before a sacrificial material is deposited, covered, and etched out leaving the channels hollow. The pit refers to a large low-pressure chamber that is in-line with the sample stream flow path that the stream comes in contact with, causing it to drop and flatten out. Simultaneously, buffer fluid is injected above the sample stream to occupy the volume above it before both rise out of the pit and they are met from all sides by additional buffer fluid provided by the trench feature. The combined fluid stream exits the junction through the outlet as a 3D focused sample stream with a form factor that is similar to the cross-section of the outlet channel and ready for optical interrogation. The fluids are typically drawn through the chip by applying vacuum pressure to the outlet fluid port, but fluid can alternatively be pushed through by applying pressure to the inlet ports.

Figure 3 shows a top down view of the optofluidic chip with the 3DHDF element integrated. The silicon-based chip is 1×1 cm^2. Four fluid reservoirs manage the sample, buffer, and waste fluid on the chip. The fluid on the chip flows from left to right, with the buffer fluid coming in contact with the sample fluid near the center of the chip at the focusing junction. The fluid focusing occurs before the liquid-core waveguide fluid channel is intersected by solid-core waveguides that carry light for fluorescence excitation and signal collection. This intersection point is called the excitation region, or excitation volume when referring to the fluid.

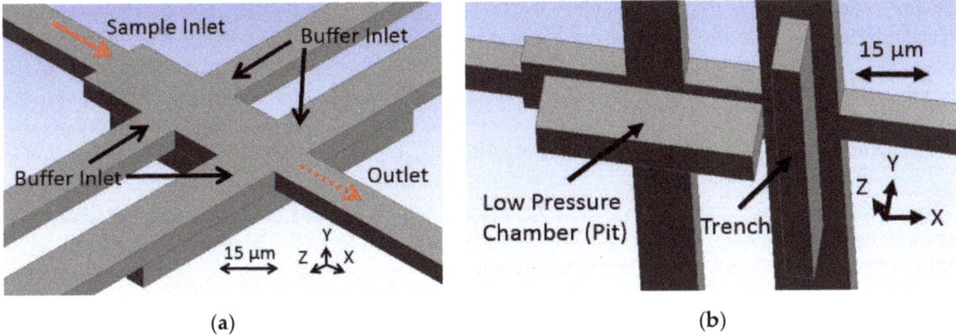

Figure 2. Schematics show (**a**) an above oblique view of the focusing junction describing the operation principle in which sample flow is shown in red, buffer flow in black, and outlet flow in dashed red, and (**b**) a beneath oblique view showing the etched features.

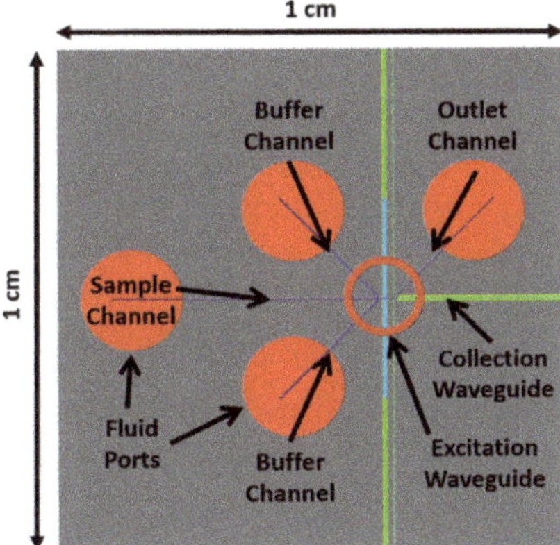

Figure 3. The focusing junction integrates into the optofluidic chip design near the center of the chip, indicated with the red ring, drawn from a top view.

2.2. Channel Dimension Determination

The fluid channel dimensions between the focusing junction at chip center and the fluid ports were determined by solving a pair of volumetric flow ratios (Q/Q) while using the known hydrodynamic relation $P = QR$ or $Q = P/R$, where P is pressure, Q is volumetric flow rate, and R is fluid resistance. This relation is derived from Newton's second law, to the Navier–Stokes equations, to the Hagen–Poiseuille relation. The relation assumes incompressible laminar Newtonian fluid flow through a circular pipe. The hydraulic radius is substituted to apply the relation to a rectangular duct. The pair of volumetric flow ratios solved were of the buffer fluid flow and sample fluid flow, and of the buffer fluid flow and the outlet fluid flow. The pressure drop was chosen to be approximately equal between the ports and the fluid focusing junction, thus canceling out and leaving just the R terms. The R terms were expanded into the Hagen–Poiseuille form, the constants canceled out, and the hydraulic radius substituted, leaving

$$\frac{Q_2}{Q_1} = \frac{L_1}{L_2}\frac{R_{H2}}{R_{H1}} = \frac{L_1}{L_2}\frac{(2w_1 + 2h_1)^4}{w_1^4 h_1^4}\frac{w_2^4 h_2^4}{(2w_2 + 2h_2)^4} \tag{1}$$

L is fluid channel length, w is fluid channel width, and h is fluid channel height. Subscript 2 indicates terms relating to the buffer fluid channel and subscript 1 indicates terms that relate to the sample (or outlet) fluid channel. This equation was input into an online graphical calculator, called Desmos (San Francisco, CA, USA), which allowed for us to keep track of parameters and view the solution visually. Each Q ratio value was input manually as calculated from the input velocities and cross-sectional areas of the inlets from the CAD model ($Q = VA$). The ratio was subtracted from both sides of the equation to create the slope intercept form for graphing. One of the dimension variables in the equation was replaced with the variable x to act as the solution. The equation was multiplied by a large number in order to cause the graphed line to appear nearly vertical for ease of dimension determination. The variable values and solution variable were adjusted until the solutions were deemed as acceptable for fabrication tolerances.

The height of the microchannels is 6 μm. The etch depth is 12 μm. The sample inlet channel is 12 μm wide and 4700 μm long. The buffer channels are 2500 μm long and 20 μm wide, but they split in two halfway from the port to the junction; the pit directed channel is 7 μm wide and the trench channel 12 μm wide. The outlet channel is 12 μm wide and the length is 2500 μm. Determining the dimensions based on volumetric flow ratio means the device can be operated by negative pressure at the outlet over a range of flow velocities.

2.3. Modeling

The design that is outlined in Figure 2 was developed in the computational fluid dynamics software Fluent (ANSYS, Canonsburg, PA, USA). The design was meant to be integrated with previously developed optofluidic channels, so it was constrained in outlet channel cross sectional dimensions of 12 μm × 6 μm [32]. The pit and trench etch features were designed to have the same depth of 12 μm so as to simplify etching processes during fabrication. Note that the pit determines the vertical dimension of the focused sample stream and the trench determines the vertical location of the focused sample stream at the outlet.

The model, as visualized in Fluent, is shown in Figure 4, where sample fluid begins as uniformly distributed particle traces and the buffer fluid is invisible. Sample fluid flows in the X direction from left to right, first expanding laterally before coming in contact with the low-pressure chamber, or pit. At this location, buffer fluid travels in the Z direction toward the center and it comes in contact with the sample fluid, pressing it down and occupying the top space of the channel. The sample stream rises up out of the low-pressure chamber before coming in contact with additional buffer fluid in the trench, also traveling in the Z direction toward the center. This raises the sample stream from the bottom toward the center of the channel, effectively 3D focusing the sample stream, as seen at the outlet (see the inset of Figure 4a). Note that the sample stream exhibits a nose shaped cross-section, where a majority of particle streams are predicted to be contained in the center of the channel vertically, but a small portion extend over the main body toward the top of the channel. This is a result of optimizing the design geometry while using large finite elements. This caused streamlines to end abruptly inside the model, resulting in an incomplete stream profile that appeared to be better focused. Increasing the element resolution (decreasing element size) overcame this issue and resulted in the complete profile seen in the figure. The model in Figure 4 used a body fitted cartesian mesh method with half micron element size, resulting in 182,222 finite elements. Although the design was optimized, tooled, and fabricated before the incomplete profile was discovered, it will be shown that the device operation could be adapted for more optimal stream focusing.

The color visible in the Fluent model represents fluid velocity of the sample stream. Blue shows slower flow streams and red faster. The sample stream flow velocity increases through the fluid junction, because the buffer fluid occupies volume and both exit the junction together. The fluid must flow faster to move the combined fluids out with the same fluid volume flow rate with which the fluids are entering the junction.

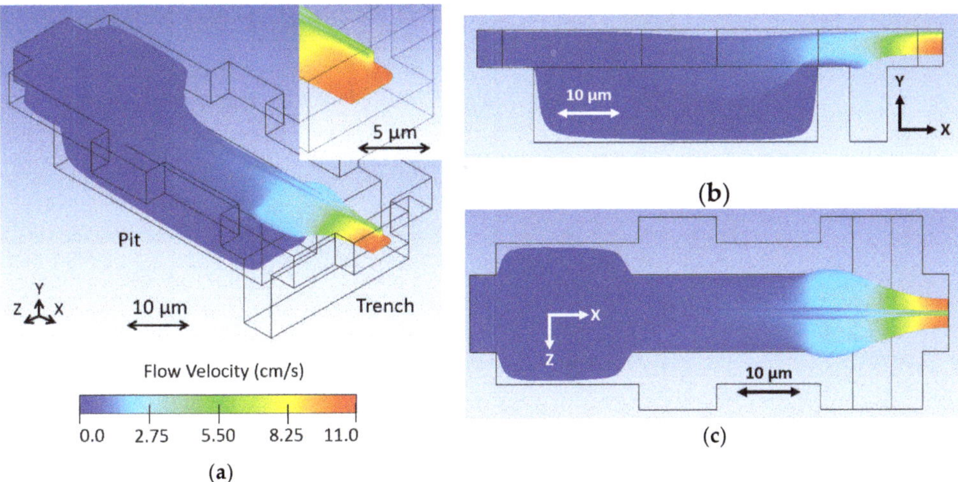

Figure 4. Sample fluid flows in the X direction of the fluid junction model in an (**a**) oblique view, (**b**) side view, and (**c**) top view, where buffer fluid is transparent and sample fluid is colored from dark blue to red, representing the range of flow velocities from slow to fast. A uniformly distributed particle concentration at the sample inlet becomes a focused particle distribution at the outlet.

The simulation was performed with sample and buffer inlet velocities of 1 cm/s and an outlet pressure of 0 Pascals. Fluid dynamics of the microchannels dictate laminar flow, thus the shortened channels in the simulation. We expect focused flow to propagate down the channel for optical interrogation. However, if the particles are diffusing outside of the focused stream, this behavior would not be accounted for by the fluid dynamic software, as seen in hundred-micron channel length models.

That the modeling conditions do not resemble the experimental conditions was a conscious choice, because the channel length is more than two orders of magnitude greater than the cross-section. Modeling the entire length of the fluid channels from inlet reservoirs to outlet reservoirs, especially given the negligible inertial focusing effects, is unhelpful. Instead, the focusing junction was modeled with the inlet velocities and outlet pressure and the channel dimensions on chip were determined by volumetric flow ratio to achieve the modeled velocities. This was expected to result in the predicted hydrodynamic effect by using negative pressure-driven flow. The experimental results show focused stream flow velocities on the same order of magnitude as predicted.

We modeled the design with some variations to portray how certain changes to the design affect the hydrodynamic focusing. First, we removed the pit feature in Figure 5a to show how it functions for vertical focusing. Without the pit, strong horizontal focusing occurs as the buffer fluid squeezes the sample stream from the sides. Subsequently, the trench feature raises up and further squeezes the sample stream. The resulting sample stream is what might be called 2.5D focused, as it is, in fact, limited in position in the fluid channel, but only focused on three sides instead of four. The pit is necessary to move the sample stream downward as buffer fluid fills the top of the channel, as described earlier.

We investigated how controlling fluid flow might enable us to overcome the suboptimal nose shaped sample stream seen in Figure 4. By increasing the buffer fluid input velocity, we expected a greater amount of vertical focusing as a result of the sample stream being more fully covered in the pit region. Figure 5b shows two and a half times the velocity, 2.5 cm/s, where the bridge of the nose shape disappears. Note that sample flow velocity increases with an increasing buffer flow, as all of the fluid entering the junction must leave through one port at the same volumetric flow rate that it enters at the other five ports combined. The increase in buffer fluid limits the amount of sample fluid that can be processed through the device per time, which should be considered in any time-sensitive optofluidic application, for instance, when screening an entire volume of liquid, which contains particles of interest. However, the increase in fluid velocity might compensate for this.

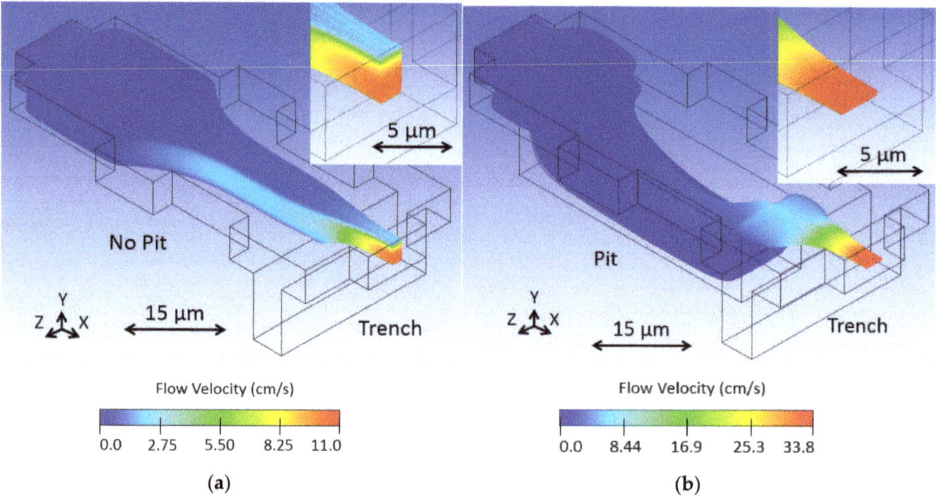

Figure 5. Variations were modeled to show how fluid channel shape and function alter the hydrodynamic focusing effect. (**a**) No pit leaves the horizontally focused sample stream near the top of the channel, a sort of 2.5D HDF. (**b**) Increasing buffer flow by 2.5 times better shapes the sample stream and removes the bridge of the nose shape.

The current design could be adapted to induce optimal flows by increasing the volumetric flow ratio of the buffer fluid. This could be performed with fluid inlet velocity control rather than negative pressure driven flow at the outlet, or the straightforward addition of back pressure at the buffer inlets while using gravity or some other means of positive pressure to push more buffer fluid through [33]. However, high operational vacuum pressures dominate such gravity-driven back pressure effects, as will be discussed.

2.4. Fabrication

The optofluidic device was fabricated in a class 10 cleanroom while using standard silicon-based microfabrication processes. Figure 6 shows side view drawings of the critical process steps. First, commercial layers of silicon dioxide and tantalum pentoxide were deposited on a blank silicon wafer at thicknesses of 265 nm and 102 nm, respectively. These dielectric layers form the anti-resonant reflecting optical waveguide structure (ARROW) required for waveguiding in a low refractive index medium such as a liquid. Next, chrome was deposited on the wafer and patterned with AZ 3330 photoresist and chrome etchant to form a stop etch feature for a later step. Next, nickel was deposited on the wafer and again patterned with AZ 3330 and nickel etchant. The features were ICP-RIE etched, forming the low-pressure chamber (pit) and trench to 12 μm deep. Here, the sacrificial material of SU-8 10 photoresist was deposited and patterned at 3000 rpm spin speed and 30 s exposure time, filling the etched features and forming the liquid-core channel volume. Note that the 6 μm tall resist overlaps all of the edges of the etched features by about 5 μm to ensure coverage and allow for some fabrication tolerance. Next, a self-aligned pedestal was created by depositing and developing AZ 4620 off the top of the SU-8 resist, then depositing nickel before removing the remaining resist with acetone. This liftoff step leaves the nickel mask on top of the defined pedestal to act as a dry etch mask for a 6 μm pedestal. The oxide was deposited next as low stress, high refractive index plasma-enhanced chemical vapor deposited (PECVD) silicon dioxide. Nickel was deposited again, patterned, and the oxide was then etched to form the liquid-core walls and solid-core bodies. An additional low stress, yet low refractive index PECVD silicon dioxide layer was deposited as an environmental barrier [34]. Finally, the sacrificial material was exposed at the ends of the channels with buffered hydrofluoric acid while using AZ 4620 as a mask, and the SU-8 was etched out with a mixture of hydrogen peroxide and sulfuric acid at a ratio of 3:2.

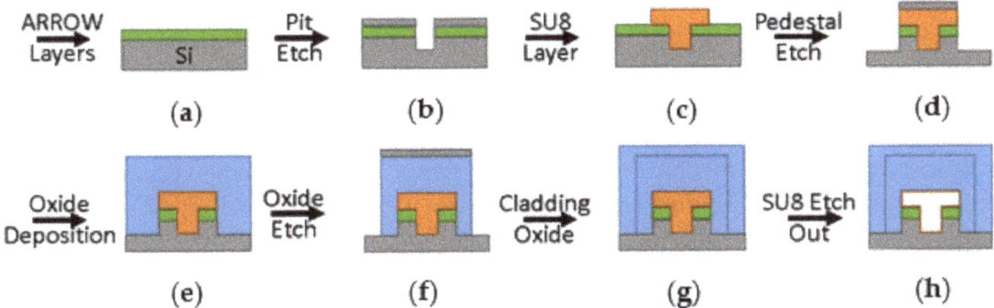

Figure 6. The fabrication flow chart shows cross-section drawing on critical silicon-based microfabrication processes used to build the device, with silicon substrate in gray, thin metal etch masks in gray, anti-resonant reflecting optical waveguide structure (ARROW) layers in green, SU-8 photoresist in orange, and PECVD oxide in blue. (**a**) ARROW layers enable liquid-core waveguiding, and (**b**) the low-pressure chamber (pit) and trench features are etched before (**c**) SU-8 photoresist forms the fluid volume. (**d**) A pedestal helps to improve yield and isolate the optics and (**e**) PECVD oxide forms the walls. (**f**) The liquid-core and solid-core waveguides are etched and (**g**) cladding PECVD oxide protects it all. (**h**) Finally, acid removes the sacrificial SU-8 photoresist, leaving the channels hollow.

The critical features of a completed device are seen from the top down in Figure 7, showing the fluid focusing junction, as well as the liquid-core channel being intersected by solid-core waveguides. Note how the solid-core features in Figure 7a rest on a pedestal wider than themselves, while the pedestal of the liquid-core channels is not visible from the top, because it only lies directly underneath these features. The pedestal strengthens the hollow channels as it moves the crevice of the oxide all the way to the etched substrate surface. Figure 7b,c show the top of the focusing junction oxide as well as the pit and trench features inside the focusing junction.

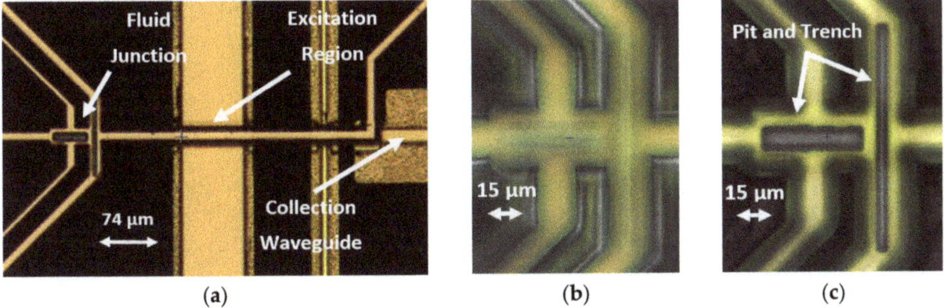

Figure 7. Optical three-dimensional (3D) profilometer images of a completed device center show (**a**) the liquid-core focusing junction and channel and the intersecting solid-core waveguides as well as (**b**) the top of the focusing junction oxide and (**c**) the inside of the focusing junction showing the etch features.

3. Results

3.1. Experimental Methods

The sample and buffer fluid flow from left to right and hydrodynamic focusing occurs in the fluid junction shortly before the optical excitation region as seen from top down in Figure 8, which correlates to the center of the chip layout found in Figure 3. It is at this excitation region where the liquid-core waveguide fluid channel and the solid-core MMI waveguide intersect and the light-matter interaction occurs by multi-spot excitation (see Figure 8b), which enables the direct determination of the particles' flow speed. The solid-core waveguides are oriented perpendicular to the edge of the chip, with the excitation waveguides vertical and the collection waveguide horizontal. Note how the liquid-core waveguide guides fluorescent signal emission to the collection waveguide, which is coupled to the liquid-core and guides the signal to the chip edge, where it can be collected and analyzed.

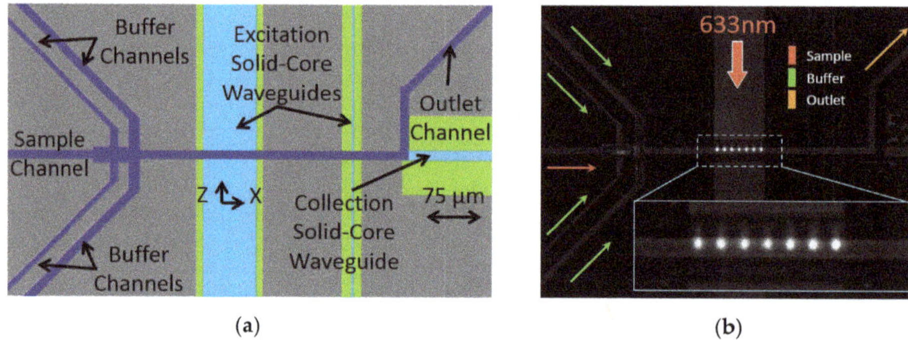

Figure 8. (a) Top down schematic of the center of the optofluidic chip. Fluid flows from left to right in the dark blue liquid-core channels, passing through the focusing junction before interacting with the light delivered by the light blue solid-core waveguides. (b) Photograph of on-chip multi-mode interference excitation waveguide (MMI) multi-spot excitation scheme using fluorescent dye and DI water.

Three-dimensional focusing was confirmed and the diffusion of the focused sample stream in buffer fluid was characterized in a previously published design similar to this one [31]. The present design shares liquid-core waveguide dimensions (12 µm × 6 µm cross-section), functional method (buffer fluid focusing), and operational flow rates (~3 cm/s), as well as modeling parameters, fabrication processes and materials, and testing methods. Good agreement between predicted and actual streamlines found in the previous design can thus translate to the current design. Moreover, the inertial focusing effects are found to be negligible by calculating the maximum Reynolds number to be less than one. The low diffusivity of the fluorescent beads in conjunction with the predicted laminar flow stream, previous experimentation, and calculated low Reynolds number make us confident that particles remain in hydrodynamically focused streams. Focusing is indicated by an analysis of signal distributions rather than repeating previous experiments.

The previous experiments were performed with Cy5 fluorescent dye and water which were drawn through the device with negative pressure and illuminated with a 633 nm wavelength laser coupled to the chip through a single mode optical fiber (Newport, Irvine, CA, USA, FS-V). The photonic signal passed through a penta-bandpass optical filter (Semrock, Rochester, NY, USA, FF01-440/521/607/694/809-25) before it was detected by a Time-Correlated Single Photon Counting (TCSPC) avalanche photodiode (APD) (PicoQuant, Berlin, Germany, TimeHarp 260 Nano) and analyzed using our event detector scripts developed in Python and available on github (San Francisco, CA, USA) (github.com/vganjali/EventDetector). It utilizes continuous wavelet transform (CWT) with custom made wavelets to enhance the signal in both time and scale domains, followed by the threshold method to localized events accurately in time and scale. Scale information is designed to match with the Δt value that was related to subsequent bright spots for a given multi-peak signal (representing the multi-spot illumination pattern shown in Figure 8b) and it is used as a means to determine the velocity of the detected fluorescent beads [35]. If using diffusible samples such as dye is desired, design constraints should be considered. This includes locating the focusing junction nearer the excitation region and flowing the sample fluid at the highest possible velocity (negative vacuum pressure) to mitigate the lateral spreading of the focused stream.

The same apparatus, as shown in Figure 9, was used for evaluating the present design, but here we used 200 nm dark red fluorescent beads (FluoSpheres™ Carboxylate-Modified Microspheres from Invitrogen™, Carlsbad, CA, USA) diluted to 10^7/mL concentration to elicit discrete signal peaks for analysis. The laser light is coupled to the multi-mode interference excitation waveguide (MMI), which was designed to generate a 75 µm long multi-spot illumination pattern at the intersecting liquid-core waveguide, as seen in Figure 8b.

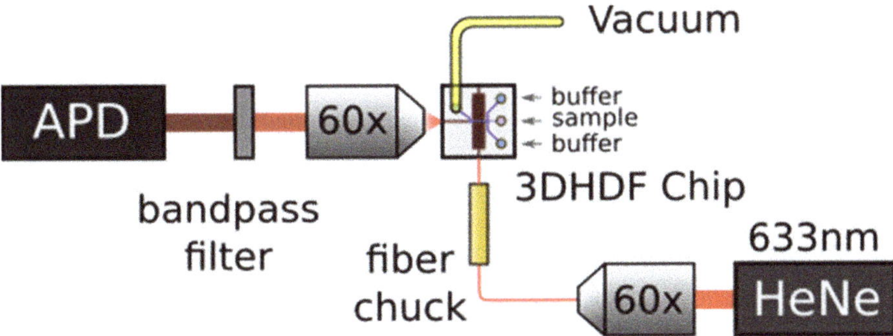

Figure 9. The test system includes the optofluidic chip connected to a vacuum for generating negative pressure to pull sample and buffer fluid through. A 633 nm laser is coupled to the excitation waveguide through an optical fiber to elicit fluorescent photon emission which is collected and detected by an avalanche photodiode (APD) for analysis.

The lateral spreading of molecules in a focused stream is approximately proportional to the square root of diffusivity. The beads have a reported diffusivity of about 2×10^{-8} cm^2/s, two orders of magnitude lower than the dye [36,37]. Two-hundred nm beads should exhibit little lateral spreading due to diffusion and be useful in characterizing the detection enhancement of the 3DHDF design. This means that the beads are representative of molecules with similar or even higher diffusivity, such as DNA, which are of interest in optofluidic interrogation.

3.2. Fluorescent Signal Coefficients of Variance

The effect of 3DHDF on fluorescent signal quality was measured by first filling all of the fluid channels with a fluorescent bead solution as a control test. In this case, there is no effective focusing of a particle stream because the entire cross section of a fluid channel is filled with particles. Measured fluorescence for this case was then compared to the case, where the sample fluid channel was filled with solution containing beads and the buffer channels with water. In both cases, fluid was drawn through the optofluidic channels with negative pressure at the outlet and illuminated with multiple spots of 633 nm laser light through the solid-core MMI waveguide. An avalanche photodiode at the chip edge collected the fluorescent signal captured orthogonally in the liquid-core waveguide and a computer reported the intensity in counts per 0.1 ms, as observed in the traces in Figure 10. The control intensity trace in Figure 10a represents a uniformly bead filled fluid channel and Figure 10b represents a focused bead stream that was sheathed by buffer fluid.

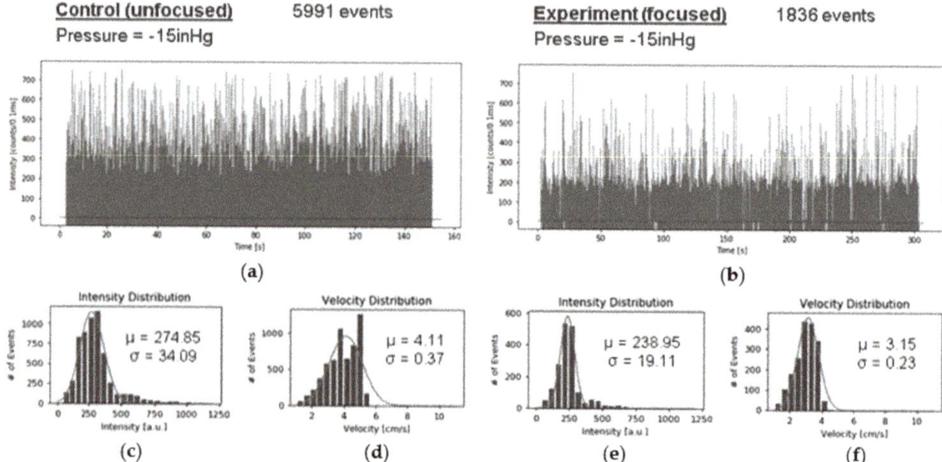

Figure 10. Signal intensity traces in counts per 0.1 ms were collected for (**a**) unfocused operation over 150 s and (**b**) focused operation over 300 s. (**c**–**f**) Histogram distributions for signal intensity and fluid velocity show more consistent signal with three-dimensional hydrodynamic focusing (3DHDF).

These plots show the measured optical intensity versus time, with intensity spikes representing fluorescing beads passing through an excitation point. When comparing the control trace to the experiment trace we see an almost sevenfold decrease in events captured per time. This is likely due to the buffer fluid in the 3DHDF case occupying a large fraction of the channel volume and reducing the number of beads passing through the channel per time. Figure 10c–f show histogram distributions of both signal intensity and fluid velocity for the control and 3DHDF cases. For both the intensity and velocity distributions, we see a narrowing due to hydrodynamic focusing, which is expected.

An improvement in detection performance can be represented by the coefficient of variance (CV) for the measured signal. This is calculated by dividing the standard deviation of the signal intensity by the mean of the distribution. A smaller CV represents a narrowing of the distribution or an increase in mean. They both indicate enhanced detection of optofluidic sensors promised by HDF. The data found in Figure 10c–f were used to generate CV values for characterizing detection enhancement. The resulting signal intensity CVs are 0.12 and 0.08 for the control and experiment, respectively, a decrease of 33%, which is one way to quantify the increase in signal uniformity that is afforded by 3DHDF. The calculated CV for the fluid velocity distributions are 0.09 and 0.07 for the control and experiment respectively, a decrease of 22%, again with 3DHDF showing improved uniformity. The data were obtained while using −15 inHg vacuum-pressure-driven flow to overcome the reservoir effects while maintaining high signal intensity, as will be discussed shortly.

3.3. Pressure Dependent Detection Enhancement

We wanted to test how the mean and standard deviation of signal traces that were collected at various vacuum pressures would relate. Increasing the vacuum pressure results in a greater pulling force on the fluid through the channels, and thus higher fluid velocities and a lower chance of detecting all emitted photons. This is related to the time resolution and dead time of the detector. Fluorescence happens on the nanosecond scale, which means the beads should have enough time to be excited and emit fluorescent signal, but the detector is unable to detect all of the received photons in the short time interval that the fluorophore spends in the excitation volume. The result is lower signal intensities. We changed the pressure in increments of 5 from −5 inHg to −25 inHg as well as −28 inHg, the limit of the vacuum system. Figure 11a shows the normalized mean and standard deviation values of the signal intensity and fluid velocity. As expected, the velocity increased and the intensity decreased with increasing vacuum pressure magnitude. The mean fluid velocity ranged from approximately 1 cm/s to 5.4 cm/s.

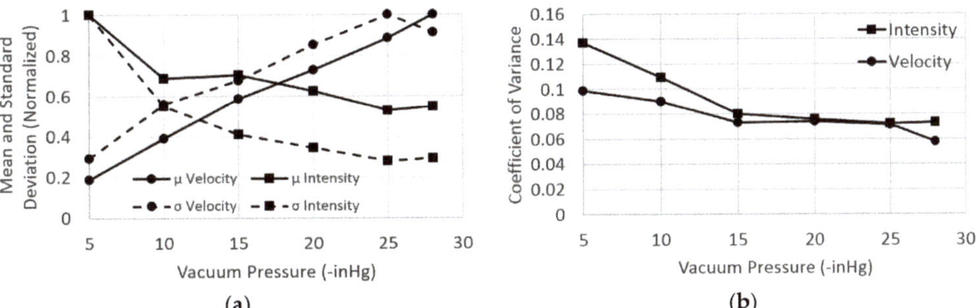

Figure 11. (**a**) With increasing vacuum pressure, the mean (μ) and standard deviation (σ) of the velocity increase while those of the intensity decrease (**b**) The coefficient of variance (CV) decreases with increasing vacuum pressure for both signal intensity and particle velocity before flattening out. Back pressure widens the focused sample stream at low operational pressures but is dominated at higher pressures.

Figure 11b plots the coefficient of variance for intensity and velocity for each vacuum pressure. The trend observed is a result of reservoir effects. This is seen as higher CV values at low operational vacuum pressures that drop with increasing vacuum pressure before flattening out. The sample channel has higher fluidic resistance than the buffer channels, due to the sample channel being narrower than the buffer channels. This means the buffer channels consumed reservoir fluid volume faster than the sample channel, which results in an increase in back pressure on the sample fluid due to the greater height of fluid in the column [33]. During experimentation, an unbalance of volumetric flow ratio occurred, resulting in a less tightly focused sample stream (higher CV). The reservoir effect is dominant at low pressures, but it becomes negligible at higher vacuum pressures, being seen as the flattening of the CV line. The flat portion matches expectations as the device is designed to operate over a range of velocities with negligible focusing change.

We concluded that a minimum of −15 inHg vacuum pressure should be used to overcome the reservoir effects that were observed in the testing of the device. This finding, in conjunction with the signal intensity data found in Figure 11a, led us to perform device characterization at −15 inHg vacuum pressure (as shown in Figure 10) to overcome the observed reservoir effects while maintaining the greatest signal intensity. Negative pressure driven flow schemes used for optofluidic detection should similarly balance signal variance with signal intensity to achieve optimal detection schemes.

We compared the device yield of the silicon-based microchips in over 300 total devices. The present design showed an average of 45% yield when compared to the average 29% yield of the previous design, a 1.56 times improvement in device yield. We attribute this yield improvement to the robust design while using a single-layer of photoresist to form the fluid channels.

4. Conclusions

A 10-μm scale, single layer 3D hydrodynamic focusing design was implemented, which improved device yield by 1.56 times over previous work. The low-pressure chamber, or pit, is the key feature that induces 3D focusing using buffer fluid that is delivered by negative pressure-driven flow, controlling the position of target particles in the fluid channel during fluorescent signal generation. This mitigates the disadvantageous natural phenomena that are present in microfluidic channels by narrowing signal intensity and flow velocity distributions and constraining target particles to highest collection efficiency regions. It does this while using just four fluid reservoirs. The concept, modeling, fabrication, and testing were outlined. Fluorescent beads representing molecules of interest were analyzed to find 33% more consistent signal with focusing. The reservoir effects were observed at low operational vacuum pressures and an optimal detection scheme was used to maximize signal consistency while maintaining high signal intensity. Signal intensity and flow velocity coefficients of variance matched expectations by decreasing between the control and the experiment, indicating successful focusing and detection enhancement. Further improvement and optimization in signal intensity can be achieved by aligning focused particle streams with the optical excitation mode center and with collection modes in the liquid-core waveguide. This could be achieved with the current geometry while using independent fluid inlet controls to flow additional buffer fluid volume. A gravity driven back-pressure method would be ineffective at higher operational vacuum pressures as the effect becomes dominated. Alternatively, the pit approach could be redesigned to efficiently operate with negative pressure-driven flow.

Author Contributions: Conceptualization, E.S.H., J.G.W., and A.R.H.; methodology, E.S.H.; validation, E.S.H., V.G., and H.S.; formal analysis, E.S.H., V.G., and H.S.; investigation, E.S.H. and V.G.; resources, A.R.H. and H.S.; data curation, V.G.; writing—original draft preparation, E.S.H.; writing—review and editing, E.S.H.; visualization, E.S.H.; supervision, A.R.H.; project administration, A.R.H. and H.S.; funding acquisition, A.R.H. and H.S. All authors have read and agreed to the published version of the manuscript.

Funding: This research was funded by the National Institutes of Health, grant numbers 1R01AI116989 and 1R01EB028608, and by the National Science Foundation, grant number CBET-1703058. The APC was funded by the National Institutes of Health, grant number 1R01AI116989.

Conflicts of Interest: A.R.H. and H.S. have a financial interest in Fluxus Inc., which commercializes optofluidic technology.

References

1. Ozcelik, D.; Cai, H.; Leake, K.D.; Hawkins, A.R.; Schmidt, H. Optofluidic bioanalysis: Fundamentals and applications. *J. Nanophotonics* **2017**, *6*, 647–661. [CrossRef] [PubMed]
2. Wu, T.-F.; Cho, S.H.; Chiu, Y.-J.; Lo, Y.-H. Lab-on-a-chip device and system for point-of-care applications. In *Handbook of Photonics for Biomedical Engineering*; Ho, A.P., Kim, D., Somekh, M., Eds.; Springer: Dordrecht, The Netherlands, 2014.
3. Wolfe, D.B.; Conroy, R.S.; Garstecki, P.; Mayers, B.T.; Fischbach, M.A.; Paul, K.E.; Mara, P.; Whitesides, G.M. Dynamic control of liquid-core/liquid-cladding optical waveguides. *PNAs* **2004**, *101*, 12434–12438. [CrossRef] [PubMed]
4. Manor, R.; Datta, A.; Ahmad, I.; Holtz, M.; Gangopadhyay, S.; Dallas, T. Microfabrication and characterization of liquid core waveguide glass channels coated with teflon af. *IEEE Sens. J.* **2003**, *3*, 687–692. [CrossRef]
5. Pascoa, R.N.M.J.; Toth, I.V.; Rangel, A.O.S.S. Review on recent applications of the liquid waveguide capillary cell in flow based analysis techniques to enhance the sensitivity of spectroscopic detection methods. *Anal. Chim. Acta* **2012**, *739*, 1–13. [CrossRef]
6. Testa, G.; Bernini, R. Integrated tunable liquid optical fiber. *Lab Chip* **2012**, *12*, 3670–3672. [CrossRef]
7. Testa, G.; Persichetti, G.; Bernini, R. Optofluidic approaches for enhanced microsensor performances. *Sensors* **2015**, *15*, 465–484. [CrossRef] [PubMed]
8. Stambaugh, A.; Meena, G.G.; Parks, J.W.; Schmidt, H.; Stott, M.A.; Hawkins, A.R. Optofluidic detection of zika nucleic acid and protein biomarkers using multimode interference multiplexing. *Biomed. Opt. Express* **2018**, *9*, 3725–3730. [CrossRef]
9. Ozcelik, D.; Parks, J.W.; Wall, T.A.; Stott, M.A.; Cai, H.; Hawkins, A.R.; Schmidt, H. Optofluidic wavelength division multiplexing for single-virus detection. *PNAs* **2015**, *112*, 12933–12937. [CrossRef]
10. Szydzik, C.; Gavela, A.F.; Herranz, S.; Roccisano, J.M.K.; Thurgood, P.; Khoshmanesh, K.; Mitchell, A.; Lechuga, L.M. An automated optofluidic biosensor platform combining interferometric sensors and injection moulded microfluidics. *Lab Chip* **2017**, *17*, 2793–2804. [CrossRef]
11. Liu, S.; Zhao, Y.; Parks, J.W.; Deamer, D.W.; Hawkins, A.R.; Schmidt, H. Correlated electrical and optical analysis of single nanoparticles and biomolecules on a nanopore-gated optofluidic chip. *Nano Lett.* **2014**, *14*, 4816–4820. [CrossRef]

12. Du, K.; Cai, H.; Park, M.; Wall, T.A.; Stott, M.A.; Alfson, K.J.; Griffiths, A.; Carrion, R.; Patterson, J.L.; Hawkins, A.R.; et al. Multiplexed efficient on-chip sample preparation and sensitive amplification-free detection of ebola virus. *Biosens. Bioelectron.* **2017**, *91*, 489–496. [CrossRef] [PubMed]

13. Cai, H.; Parks, J.W.; Wall, T.A.; Stott, M.A.; Stambaugh, A.; Alfson, K.; Griffiths, A.; Mathies, R.A.; Carrion, R.; Patterson, J.L.; et al. Optofluidic analysis system for amplification-free, direct detection of ebola infection. *Sci. Rep.* **2015**, *5*, 14494. [CrossRef] [PubMed]

14. Fan, X.; White, I.M. Optofluidic microsystems for chemical and biological analysis. *Nat. Photonics* **2011**, *5*, 591–597. [CrossRef] [PubMed]

15. Risk, W.P.; Kim, H.-C.; Miller, R.D.; Temkin, H.; Gangopadhyay, S. Optical waveguides with an aqueous core and a low-index nanoporous cladding. *Opt. Express* **2004**, *12*, 6446–6455. [CrossRef] [PubMed]

16. Yin, D.; Schmidt, H.; Barber, J.P.; Hawkins, A.R. Integrated arrow waveguides with hollow cores. *Opt. Express* **2004**, *12*, 2710–2715. [CrossRef]

17. Black, J.A.; Hamilton, E.; Hueros, R.A.R.; Parks, J.W.; Hawkins, A.R.; Schmidt, H. Enhanced detection of single viruses on-chip via hydrodynamic focusing. *IEEE J. Sel. Top. Quantum Electron.* **2019**, *25*, 1–6. [CrossRef]

18. Rosenauer, M.; Vellekoop, M.J. Characterization of a microflow cytometer with an integrated three-dimensional optofluidic lens system. *Biomicrofluidics* **2010**, *4*, 43005. [CrossRef]

19. Xiong, B.; Wang, L.; Wang, Y.; Bao, Y.; Jiang, S.; Ye, M. Three-dimensional hydrodynamic focusing microfluidic emitter: A strategy to inhibit sample ion expansion in nanoelectrospray ionization. *Analyst* **2016**, *141*, 177–182. [CrossRef]

20. Zhao, J.; You, Z. Microfluidic hydrodynamic focusing for high-throughput applications. *J. Micromech. Microeng.* **2015**, *25*, 12. [CrossRef]

21. Zhao, Y.; Li, Q.; Hu, X. Universally applicable three-dimensional hydrodynamic focusing in a single-layer channel for single cell analysis. *Anal. Methods* **2018**, *10*, 3489–3497. [CrossRef]

22. Chiu, Y.J.; Cho, S.H.; Mei, Z.; Lien, V.; Wu, T.F.; Lo, Y.H. Universally applicable three-dimensional hydrodynamic microfluidic flow focusing. *Lab Chip* **2013**, *13*, 1803–1809. [CrossRef] [PubMed]

23. Liu, Y.; Shen, Y.; Duan, L.; Yobas, L. Two-dimensional hydrodynamic flow focusing in a microfluidic platform featuring a monolithic integrated glass micronozzle. *Appl. Phys. Lett.* **2016**, *109*, 144101. [CrossRef]

24. Yang, R.; Feeback, D.L.; Wang, W. Microfabrication and test of a three-dimensional polymer hydro-focusing unit for flow cytometry applications. *Sens. Actuator A Phys.* **2005**, *118*, 259–267. [CrossRef]

25. Sundararajan, N.; Pio, M.S.; Lee, L.P.; Berlin, A.A. Three-dimensional hydrodynamic focusing in polydimethylsiloxane (PDMS) microchannels. *J. Microelectromech. Syst.* **2004**, *13*, 559–567. [CrossRef]

26. Paje, P.; Bragheri, F.; Vazquez, R.M.; Osellame, R. Straightforward 3D hydrodynamic focusing in femtosecond laser fabricated microfluidic channels. *Lab Chip* **2014**, *14*, 1826–1833.

27. Mao, X.; Nawaz, A.A.; Lin, S.C.; Lapsley, M.I.; Zhao, Y.; McCoy, J.P.; El-Deiry, W.S.; Huang, T.J. An integrated, multiparametric flow cytometry chip using "microfluidic drifting" based three-dimensional hydrodynamic focusing. *Biomicrofluidics* **2012**, *6*, 24113–241139. [CrossRef]

28. Lee, M.G.; Choi, S.; Park, J.K. Three-dimensional hydrodynamic focusing with a single sheath flow in a single-layer microfluidic device. *Lab Chip* **2009**, *9*, 3155–3160. [CrossRef]

29. Stoecklein, D.; Wu, C.-Y.; Kim, D.; Carlo, D.D.; Ganapathysubramanian, B. Optimization of micropillar sequences for fluid flow sculpting. *Phys. Fluids* **2016**, *28*, 012003. [CrossRef]

30. Hamilton, E.S.; Wright, J.G.; Stott, M.A.; Hawkins, A.R.; Black, J.A.; Schmidt, H. 3D hydrodynamic focusing for optofluidics using a stacked channel design. In Proceedings of the Conference on Lasers and Electro-Optics 2018, OSA Technical Digest (online), San Jose, CA, USA, 13–18 May 2018; Optical Society of America: Washington, DC, USA, 2018.

31. Hamilton, E.S.; Ganjalizadeh, V.; Wright, J.G.; Pitt, W.G.; Schmidt, H.; Hawkins, A.R. 3D hydrodynamic focusing in microscale channels formed with two photoresist layers. *Microfluid. Nanofluid.* **2019**, *23*, 122. [CrossRef]

32. Lunt, E.J.; Wu, B.; Keeley, J.M.; Measor, P.; Schmidt, H.; Hawkins, A.R. Hollow arrow waveguides on self-aligned pedestals for improved geometry and transmission. *IEEE Photonic. Technol. Lett.* **2010**, *22*, 1147–1149. [CrossRef]

33. Gnyawali, V.; Saremi, M.; Kolios, M.C.; Tsai, S.S.H. Stable microfluidic flow focusing using hydrostatics. *Biomicrofluidics* **2017**, *11*, 034104. [CrossRef] [PubMed]

34. Wall, T.; Hammon, S.; Hamilton, E.; Zacheu, G.; Orfila, M.; Schmidt, H.; Hawkins, A.R. Mitigating water absorption in waveguides made from unannealed pecvd sio. *IEEE Photonic. Technol. Lett.* **2017**, *29*, 806–809. [CrossRef] [PubMed]

35. Ganjalizadeh, V.; Meena, G.G.; Stott, M.A.; Schmidt, H.; Hawkins, A.R. Single particle detection enhancement with wavelet-based signal processing technique. In Proceedings of the Conference on Lasers and Electro-Optics, OSA Technical Digest, San Jose, CA, USA, 5–10 May 2019; Optical Society of America: Washington, DC, USA, 2019.

36. Kamata, M.; Takaba, Y.; Taguchi, Y.; Nagasaka, Y. Rapid and label-free sensing of intermolecular interactions using compact optical diffusion sensor. *Int. J. Heat Mass Transf.* **2019**, *133*, 73–79. [CrossRef]

37. Kuyper, C.L.; Fujimoto, B.S.; Zhao, Y.; Schiro, P.G.; Chiu, D.T. Accurate sizing of nanoparticles using confocal correlation spectroscopy. *J. Phys. Chem. B* **2006**, *110*, 24433–24441. [CrossRef] [PubMed]

 *micromachines*

Communication

Paper-based Photocatalysts Immobilization without Coffee Ring Effect for Photocatalytic Water Purification

Qingwei Li, Huichao Lin, Xiaowen Huang *, Maocui Lyu, Hongxia Zhang *, Xiaoning Zhang * and Ruiming Wang

Department of Bioengineering, State Key Laboratory of Biobased Material and Green Papermaking, Advanced Research Institute for Multidisciplinary Science, School of Food Science and Engineering, Qilu University of Technology (Shandong Academy of Sciences), Jinan 250353, China; liqingwei@qlu.edu.cn (Q.L.); linhuichao2020@gmail.com (H.L.); lvmaocui@163.com (M.L.); ruiming3k@163.com (R.W.)
* Correspondence: huangxiaowen2013@gmail.com (X.H.); xiaoningzhang@126.com (H.Z.); zhanghongxia326@hotmail.com (X.Z.)

Received: 4 January 2020; Accepted: 23 February 2020; Published: 26 February 2020

Abstract: Photocatalytic water purification is important for the degradation of organic pollutants, attracting intensive interests. Photocatalysts are preferred to be immobilized on a substrate in order to reduce the laborious separation and recycling steps. To get uniform irradiation, the photocatalysts are preferred to be even/uniform on the substrate without aggregation. Generally, the "coffee ring effect" occurs on the substrate during solvent evaporation, unfortunately resulting in the aggregation of the photocatalysts. This aggregation inevitably blocks the exposure of active sites, reactant exchange, and light absorption. Here, we reported a paper-based photocatalyst immobilization method to solve the "coffee ring" problem. We also used a "drop reactor" to achieve good photocatalytic efficiency with the advantages of large surface area, short diffusion lengths, simple operation, and uniform light absorption. Compared with the coffee ring type, the paper-based method showed higher water purification efficiency, indicating its potential application value in the future.

Keywords: photocatalysis; microreactor; optofluidics; photocatalytic water purification; paper

---

1. Introduction

Water pollution is a serious problem especially in modern industrial cities, which is a bad by-product of the prosperity of industry. Photocatalytic water purification is important for the degradation of organic pollutants [1–5], attracting intensive interest and the development of various research works on photoreactors, photocatalysts, and illumination. However, the photocatalytic efficiency of the traditional slurry method is still disappointing due to the low surface to volume ratio (reported as <600 $m^2 \cdot m^{-3}$), long diffusion length, and low photon utilization [6]. To improve the photocatalytic efficiency, photoreactors should possess the advantages of fast mass transfer and photon transfer.

Optofluidic reactors have intrinsic merits, such as fine flow control, large surface-to-volume ratio, and direct delivery of light to the reaction surface, making them natural photoreactors for photocatalysis [7–10]. Some studies showed microfluidic methods for achieving good degradation efficiency of polluted water [11–14]. Zhang et al. reported a simple and efficient microfluidic chip with wedge-structure channel for methylene blue (MB) degradation under UV light irradiation [15]. Wang et al. demonstrated a microfluidic photoelectrocatalytic reactor for water purification, in which the positive and negative bias potentials were applied to suppress the electron/hole recombination [4]. Madeddine Azzouz et al. reported a rapid and effective photocatalysis microreactor which integrates in situ grown ZnO nanowires (NW) as an efficient photocatalytic nanomaterial layer. This ZnO NW layer provided an enhanced surface area, minimizing the required interaction time to produce efficient purification performance [16]. Li et al. proposed an optofluidic microreactor with staggered micropillars in the reaction microchamber to mainly enlarge the surface area for loading catalysts and increase the active surface area [17]. In these works, the photocatalysts were all immobilized in different ways because photocatalysts are preferred to be immobilized on a substrate in order to reduce the laborious separation and recycling steps [18].

In our previous work [19], we developed a drop reactor method, which showed the merits of large surface areas, short diffusion lengths, simple operation, as well as higher efficiency than the traditional slurry method. However, a serious "coffee ring effect" occurred during the evaporation process, unfortunately resulting in the aggregation of the photocatalysts, which could inevitably block the exposure of active sites, reactant exchange, and light absorption. Thus, making the photocatalysts even/uniform on the substrate without aggregation (coffee ring effect) in a drop reactor has significant potential applications.

In this work, we reported a paper-based photocatalyst immobilization method to solve the "coffee ring" problem. A mesoporous graphite carbon nitride (mpg-C_3N_4) was chosen to be the photocatalyst, because it has been proven to be a good photocatalyst material in both others' works and our previous work [9,20–25]. Experiments on photo-degradation of methylene blue were adopted to test the effect of both coffee ring and paper-based methods. With the same concentration of mpg-C_3N_4, the degradation rate of the paper-based method is much higher/faster than the coffee ring method. The color of methylene blue in the paper-based method showed obvious lightening after 6 min, which can be clearly distinguished by the naked eye. Thus, this simple and highly efficient method with the advantages of being "coffee ring effect-free" and uniform light absorption offers a convenient tool for both photocatalytic water purification and self-cleaning paper development.

2. Materials and Methods

2.1. Drop-reactor Methods: The Coffee Ring Method and Paper-based Method

The drop-reactor method is followed as conducted in our previous paper [19]. The uncross-linked polydimethylsiloxane (PDMS, 10:1) was first poured onto a glass slide. After the PDMS layer was cured, we punched wells with a thickness of ~0.35 mm and a diameter of 9 mm. The schematic illustration of the process is shown in Figure 1A,B. Mpg-C_3N_4 in ethanol (1 mg/mL, 1× and 2 mg/mL, 2×) were dispersed by ultrasound for 30 min, respectively.

Fabrication of the coffee ring (CR) type reactor (Figure 1A): 20 µL of the mpg-C_3N_4 suspension was added to the prepare hole. The ethanol was removed by evaporation at 50 °C. The left part of the reactor is mpg-C_3N_4 (20 µg or 40 µg, equal to CR-1X or CR-2X, a radius of 2.5 mm), which shows an obvious coffee ring effect. After adding the reaction solution (20 µL), a thin glass slide was placed on the well to avoid evaporation.

Fabrication of paper-based reactor (Figure 1B): the mpg-C_3N_4 suspension was filtered by a piece of qualitative medium-speed filter paper, and the concentration (g/m^2) was adjusted to the concentration used in the coffee ring type reactor. In detail, the concentration of mpg-C_3N_4 on the filter paper was calculated by the amount of mpg-C_3N_4 in ethanol over the area of the filter paper (g/m^2). Based on the natural function of the qualitative medium-speed filter paper (high porosity, small pore size), almost all mpg-C_3N_4 dispersed in the liquid could be trapped in the process of vacuum filtration. During the evaporation of ethanol, no aggregation of the materials occurs, thus resulting in uniform distribution. The reaction solution (20 µL) was dropped on the mpg-C_3N_4-paper and covered with a thin glass slide.

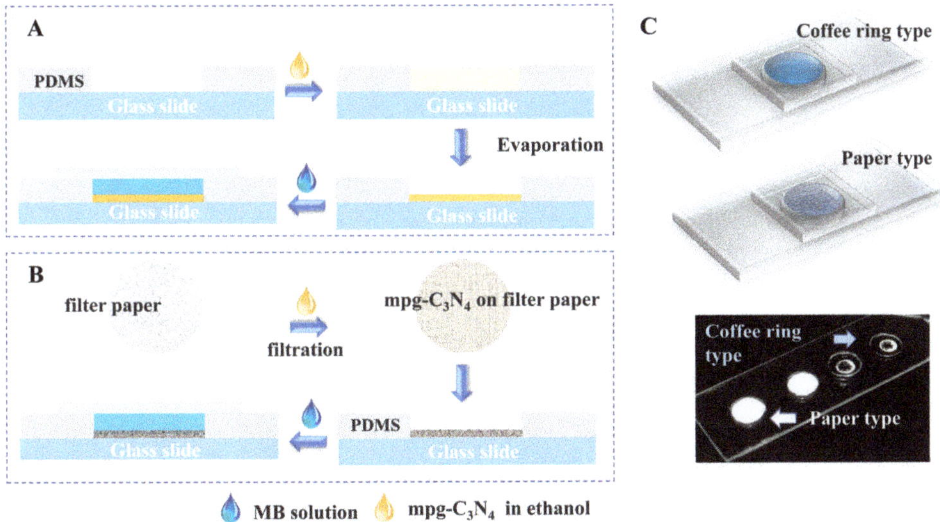

Figure 1. (**A**) Diagram of the coffee ring type drop reactor method, which occurs on the substrate during solvent evaporation, unfortunately resulting in the aggregation of the photocatalysts. (**B**) Diagram of the paper based drop reactor method. During the filtration process, photocatalysts were trapped evenly on the filter paper. (**C**) The schematic and photo of the coffee ring method and paper based method, clearly showing the inevitable aggregation of the photocatalysts in the coffee ring method.

2.2. Synthesis of mpg-C$_3$N$_4$

The mpg-C$_3$N$_4$ was synthesized as follows [26]: 5 g of cyanamide and 12.5 g of Ludox-HS 40 colloidal silica suspension (1:1 of solid ratio) were mixed until the cyanamide was completely dissolved. An oil bath at 100 °C was used to remove water. After grinding the obtained white solid with a mortar and pestle, we transferred it to a crucible and heated it with the lid on under air at 2.3 °C min^{-1} up to 550 °C (4 hours) and kept the sample at 550 °C for another 4 hours. The resultant yellow powder was treated with 4 M NH$_4$HF$_2$ solution and stirred for 48 hours. The dispersion was then filtered; the precipitate was thoroughly rinsed with deionized water and ethanol. After the filtering procedure, the yellow powder was dried under vacuum at 60 °C overnight. In this work, we used the same batch of mpg-C$_3$N$_4$ as in our former work [19], thus the characterization of mpg-C$_3$N$_4$ is the same as the former data.

2.3. Degradation of MB

A methylene blue (MB) solution of a concentration of 3×10^{-4} mol^{-1} was used as the model chemical for characterizing the photoreactivity. Degradation of MB was carried out using a Xenon lamp (300 W, PerfectLight, Beijing, China) through a 420-nm cut-off filter.

3. Results

3.1. Drop-reactor Methods: The Coffee Ring Method and Paper-based Method

As shown in Figure 1A, the photocatalysts were suspended in ethanol first and then deposited on the glass substrate during solvent evaporation. Unfortunately, this also resulted in the coffee ring effect, which means the aggregation of the photocatalysts occurred on the substrate. However, the paper based method benefited from the filtration process, in which the photocatalysts were trapped evenly on the filter paper during negative pressure filtration process (Figure 1B). In Figure 1C, the schematic and photo clearly show the differences between the two methods. To clearly understand the morphology and distribution of the photocatalysts on the filter paper, we characterized the sample by the SEM as shown in Figure S1.

3.2. Photocatalytic Degradation of MB

The prepared coffee ring method drop reactor and paper based drop reactor were used to perform the photocatalytic degradation of MB. Figure 2 shows the degradation results of the coffee ring method (CR-2×). The blue color became gradually lighter over the increase in illumination time (Figure 2A), and the UV-vis spectra (Figure 2B) correspondingly presents a clear decreasing trend. Based on the typical peak (656.5 nm) of MB, the standard curve is shown in Figure S2. The degradation rate (%) was calculated by $(C_0\text{-}C_t)/C_0$. C_0 is the concentration of MB at 0 min, C_t is the concentration of MB at t min.

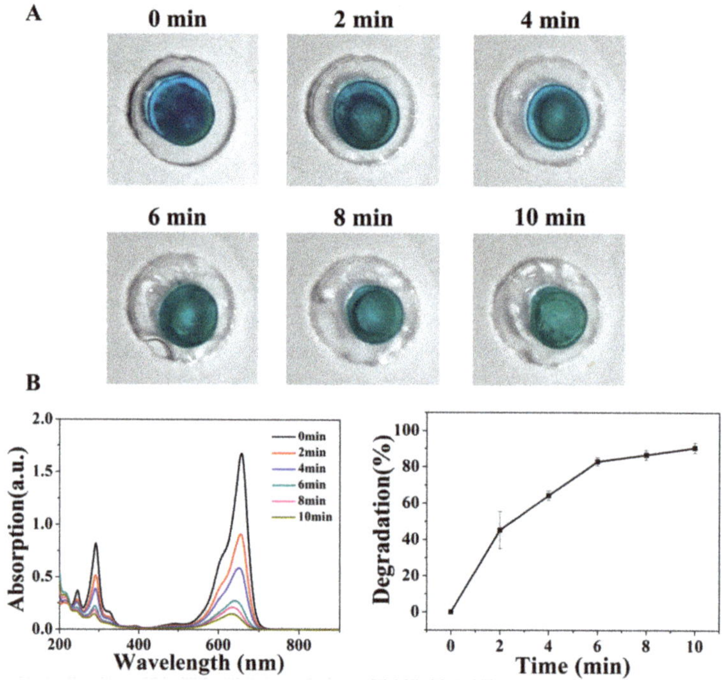

Figure 2. Degradation results of the coffee ring drop reactor method. (**A**) The blue color became gradually lighter as illumination time increased (0–10 min). (**B**) UV-vis spectra of the MB solution after illumination, and the corresponding degradation rates. Each test was repeated three times.

Figure 3 shows the degradation results of the paper based drop reactor method (2×). The blue color became obviously lighter after illumination (Figure 3A) and the corresponding UV-vis spectra (Figure 3B) present a clear decreasing trend. Compared with the coffee ring method, the paper based method showed much higher degradation results at the same time point. For example, at 4 min, the degradation rate of the coffee ring method was 64% while the degradation rate was 94% in the paper based method, and at 10 min, the degradation rater was 85% in the coffee ring method while ~100% in the paper based method.

To further explain the possible reason for the slow reaction rate in the coffee ring method, different amounts of photocatalysts were tested and the results were shown in Figure 4. The residual rate (%) was calculated by C_t/C_0. With different amounts of photocatalysts (1× or 2×), the paper based method showed better degradation ability than the coffee ring method. CR-2× showed a slightly better efficiency than CR-1×, corresponding to the aggregation of photocatalysts in the coffee ring method, which is an obvious reason for the slow reaction rate. Paper-2× showed similar results as paper-1×, indicating the saturation of the illumination site or reaction site. To get a better understanding of the reaction rate, the residual was plotted as a function of time with the exponential relationship:

$$y = A \cdot \exp(-x/t_c) + y_0 \tag{1}$$

where t_c is the characteristic chemical time, defined as the time required for the concentration of A to fall from its initial value to a value equal to 1/e times of the initial value.

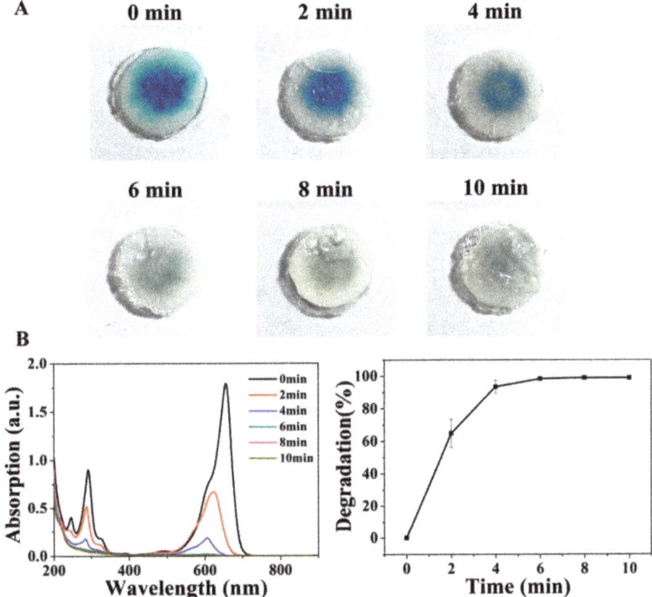

Figure 3. Degradation results of the paper based method. (**A**) The blue color obviously changed after illumination (0–10 min). (**B**) The corresponding UV-vis spectra and the degradation rate. Each test was repeated three times.

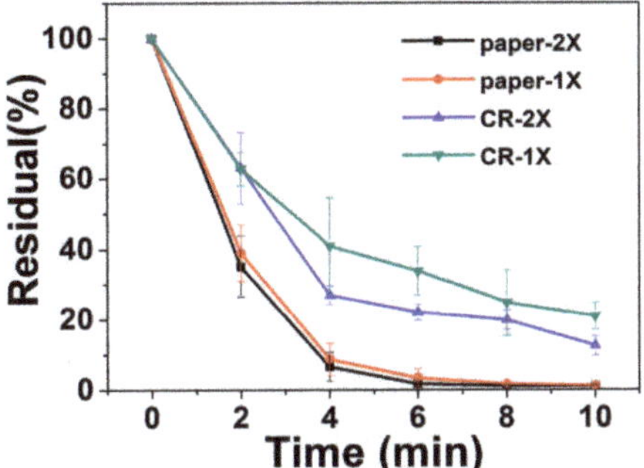

Figure 4. Comparison of the two methods with different loading amounts of mpg-C$_3$N$_4$. Degradation results of Paper type (paper-1× and paper-2×) show better efficiency than the coffee ring method (CR-1× and CR-2×). Paper-2× shows similar results to paper-1×, and CR-2× only shows a slightly better efficiency than CR-1×, corresponding to the aggregation of photocatalysts in the coffee ring method, which is an obvious reason for the slow reaction rate.

After fitting the data in the origin, we found the degradation rate of paper-1× was 1.69 times of CR-1×. CR-2× is only 1.11 times of CR-1×.

4. Discussion

In this work, the paper-immobilized photocatalysts method shows better degradation efficiency than the coffee ring method. This improvement can be attributed to the lack of aggregation and the uniform irradiation features, which cause the better exposure of active sites, reactant exchange, and light absorption.

5. Conclusions

We proposed a paper-based photocatalyst immobilization method for photocatalytic water purification, which showed better degradation efficiency than the evaporation method (due to coffee ring effect formation). The paper-based immobilization method, with the advantages of high efficiency, no aggregation, and uniform light absorption, offers a convenient tool for both photocatalytic water purification and self-cleaning paper development in the future.

Supplementary Materials: The following are available online at http://www.mdpi.com/2072-666X/11/3/244/s1, Figure S1: To clearly understand the morphology and distribution of the photocatalysts on the filter paper, we characterized the sample by the SEM, Figure S2: Standard curve of MB solution, showing the relationship between concentration and absorption.

Author Contributions: X.H., Q.L., and H.Z. conceived and designed the experiments; H.L. performed the experiments; X.Z., R.W., and M.L. contributed reagents/materials/analysis tools; X.H. wrote the paper. All authors have read and agreed to the published version of the manuscript.

Funding: This work is supported by Shandong Provincial Key Research and Development Project (2018YFJH0401), Shandong Province Natural Science Foundation (ZR2019BB006) and National Natural Science Foundation of China (21808114).

Conflicts of Interest: The authors declare no conflict of interest.

References

1. Mayes, A.M. Science and Technology for Water Purification in the Coming Decades. *Nature* **2008**, *452*, 301–310.
2. Bahnemann, D. Photocatalytic water treatment: Solar energy applications. *Sol. Energy* **2004**, *77*, 445–459. [CrossRef]
3. Nan, M.; Jin, B.; Chow, C.W.K.; Saint, C. Recent developments in photocatalytic water treatment technology: A review. *Water Res.* **2010**, *44*, 2997–3027.
4. Wang, N.; Zhang, X.; Chen, B.; Song, W.; Chan, N.Y.; Chan, H.L.W. Microfluidic photoelectrocatalytic reactors for water purification with an integrated visible-light source. *Lab Chip* **2012**, *12*, 3983–3990. [CrossRef]
5. Wang, N.; Tan, F.; Zhao, Y.; Tsoi, C.C.; Fan, X.; Yu, W.; Zhang, X. Optofluidic UV-Vis spectrophotometer for online monitoring of photocatalytic reactions. *Sci. Rep.* **2016**, *6*, 28928. [CrossRef]
6. Wang, N.; Zhang, X.; Wang, Y.; Yu, W.; Chan, H.L.W. Microfluidic reactors for photocatalytic water purification. *Lab Chip* **2014**, *14*, 1074–1082. [CrossRef]
7. Colmenares, J.C.; Rajender, S. Varma; Vaishakh Nair Selective photocatalysis of lignin-inspired chemicals by integrating hybrid nanocatalysis in microfluidic reactors. *Chem. Soc. Rev.* **2017**, *46*, 6675. [CrossRef]
8. Cheng, X.; Chen, R.; Zhu, X.; Liao, Q.; He, X.; Li, S.; Li, L. Optofluidic membrane microreactor for photocatalytic reduction of CO2. *Int. J. Hydrog. Energy* **2016**, *41*, 2457–2465. [CrossRef]
9. Huang, X.; Liu, J.; Yang, Q.; Liu, Y.; Zhu, Y.; Li, T.; Tsang, Y.H.; Zhang, X. Microfluidic chip-based one-step fabrication of an artificial photosystem I for photocatalytic cofactor regeneration. *RSC Adv.* **2016**, *6*, 101974–101980. [CrossRef]
10. Huang, X.; Wang, J.; Li, T.; Wang, J.; Xu, M.; Yu, W.; El Abed, A.; Zhang, X. Review on optofluidic microreactors for artificial photosynthesis. *Beilstein J. Nanotechnol.* **2018**, *9*, 30–41. [CrossRef] [PubMed]
11. Patinglag, L.; Sawtell, D.; Iles, A.; Melling, L.M.; Shaw, K.J. A Microfluidic Atmospheric-Pressure Plasma Reactor for Water Treatment. *Plasma Chem. Plasma Process.* **2019**, *39*, 561–575. [CrossRef]

12. Fern, J.; Garrig, G.; Cazorla, D. Photo-microfluidic chip reactors for propene complete oxidation with TiO2 photocalyst using UV-LED light. *J. Environ. Chem. Eng.* **2019**, *7*, 103408.

13. Liu, A.; Li, Z.; Wu, Z.; Xia, X. Study on the photocatalytic reaction kinetics in a TiO2 nanoparticles coated microreactor integrated microfluidics device. *Talanta* **2018**, *182*, 544–548. [CrossRef] [PubMed]

14. Lian, H.; Meng, Z. A novel and highly photocatalytic " TiO2wallpaper " made of electrospun TiO2/bioglass hybrid nanofiber. *Mater. Sci. Semicond. Process.* **2018**, *80*, 68–73. [CrossRef]

15. Zhang, H.; Wang, J.J.; Fan, J.; Fang, Q. Microfluidic chip-based analytical system for rapid screening of photocatalysts. *Talanta* **2013**, *116*, 946–950. [CrossRef] [PubMed]

16. Azzouz, I.; Habba, Y.G.; Capochichi-gnambodoe, M.; Marty, F.; Vial, J. Zinc oxide nano-enabled microfluidic reactor for water purification and its applicability to volatile organic compounds. *Microsystems Nanoeng.* **2018**, *4*, 17093. [CrossRef]

17. Li, L.; Chen, R.; Liao, Q.; Zhu, X.; Wang, G.; Wang, D. High surface area optofluidic microreactor for redox mediated photocatalytic water splitting. *Int. J. Hydrog. Energy* **2014**, *39*, 19270–19276. [CrossRef]

18. Chen, R.; Li, L.; Zhu, X.; Wang, H.; Liao, Q.; Zhang, M. Highly-durable optofluidic microreactor for photocatalytic water splitting. *Energy* **2015**, *83*, 797–804. [CrossRef]

19. Huang, X.; Hao, H.; Liu, Y.; Zhu, Y.; Zhang, X. Rapid screening of graphitic carbon nitrides for photocatalytic cofactor regeneration using a drop reactor. *Micromachines* **2017**, *8*, 175. [CrossRef]

20. Huang, J.; Antonietti, M.; Liu, J. Bio-inspired carbon nitride mesoporous spheres for artificial photosynthesis: Photocatalytic cofactor regeneration for sustainable enzymatic synthesis. *J. Mater. Chem. A* **2014**, *2*, 7686. [CrossRef]

21. Liu, J.; Liu, Y.; Liu, N.; Han, Y.; Zhang, X.; Huang, H.; Lifshitz, Y.; Lee, S.; Zhong, J.; Kang, Z. Metal-free efficient photocatalyst for stable visible water splitting via a two-electron pathway. *Science* **2015**, *347*, 970–974. [CrossRef] [PubMed]

22. Zhou, Z.; Shen, Y.; Li, Y.; Liu, A.; Liu, S.; Zhang, Y. Chemical cleavage of layered carbon nitride with enhanced photoluminescent performances and photoconduction. *ACS Nano* **2015**, *9*, 12480–12487. [CrossRef] [PubMed]

23. Wang, X.; Maeda, K.; Thomas, A.; Takanabe, K.; Xin, G.; Carlsson, J.M.; Domen, K.; Antonietti, M. A metal-free polymeric photocatalyst for hydrogen production from water under visible light. *Nat. Mater.* **2009**, *8*, 76–80. [CrossRef]

24. Liu, J.; Wang, H.; Antonietti, M. Graphitic carbon nitride "reloaded": Emerging applications beyond (photo)catalysis. *Chem. Soc. Rev.* **2016**, *45*, 2308–2326. [CrossRef]

25. Duan, J.; Chen, S.; Jaroniec, M.; Qiao, S.Z. Porous C3N4 Nanolayers@N-Graphene Films as Catalyst Electrodes for Highly Efficient Hydrogen Evolution. *ACS Nano* **2015**, *9*, 931–940. [CrossRef]

26. Liu, J.; Antonietti, M. Bio-inspired NADH regeneration by carbon nitride photocatalysis using diatom templates. *Energy Environ. Sci.* **2013**, *6*, 1486–1493. [CrossRef]

 *micromachines*

Article

Dye-Doped ZnO Microcapsules for High Throughput and Sensitive Optofluidic Micro-Thermometry

Najla Ghifari [1,2,†], **Sara Rassouk** [1,†], **Zain Hayat** [1,‡], **Abdelhafed Taleb** [3], **Adil Chahboun** [2] and **Abdel I. El Abed** [1,*]

[1]   Laboratoire de Photonique Quantique et Moléculaire (LPQM), UMR 8537, Ecole Normale Supérieure Paris Saclay, CentraleSupélec, CNRS, Université Paris-Saclay, 94235 Cachan, France; najla.ghifari@ens-paris-saclay.fr (N.G.); sara.rsok@gmail.com (S.R.); zainhayat@hotmail.com (Z.H.)

[2]   Laboratoire des Couches Minces et Nanomatériaux (CMN), FST Tanger, Université Abdelmalek Essaadi, 90000 Tangier, Morocco; adchahboun@uae.ac.ma

[3]   PSL University, Chimie ParisTech—CNRS, Institut de Recherche de Chimie Paris, Paris 75005, France; Sorbonne université, 4 place Jussieu, 75231 Paris, France; abdelhafed.taleb@upmc.fr

*   Correspondence: abdel.el-abed@ens-paris-saclay.fr

†   These authors contributed equally to this work.

‡   Present address: Microsystèmes D'Analyse (MICA), Laboratoire D'analyse et D'architecture des Systèmes (LAAS), CNRS, 31400 Toulouse, France.

Received: 7 December 2019; Accepted: 7 January 2020; Published: 17 January 2020

Abstract: The main objective of this work is to show the proof of concept of a new optofluidic method for high throughput fluorescence-based thermometry, which enables the measure of temperature inside optofluidic microsystems at the millisecond (ms) time scale (high throughput). We used droplet microfluidics to produce highly monodisperse microspheres from dispersed zinc oxide (ZnO) nanocrystals and doped them with rhodamine B (RhB) or/and rhodamine 6G (Rh6G). The fluorescence intensities of these two dyes are known to depend linearly on temperature but in two opposite manner. Their mixture enables for the construction of reference probe whose fluorescence does not depend practically on temperature. The use of zinc oxide microparticles as temperature probes in microfluidic channels has two main advantages: (i) avoid the diffusion and the adsorption of the dyes inside the walls of the microfluidic channels and (ii) enhance dissipation of the heat generated by the focused incident laser beam thanks to the high thermal conductivity of this material. Our results show that the fluorescence intensity of RhB decreases linearly with increasing temperature at a rate of about $-2.2\%/^\circ$C, in a very good agreement with the literature. In contrast, we observed for the first time a nonlinear change of the fluorescence intensity of Rh6G in ZnO microparticles with a minimum intensity at a temperature equal to 40 $^\circ$C. This behaviour is reproducible and was observed only with ZnO microparticles doped with Rh6G.

Keywords: micro-thermometry; laser induced fluorescence; droplet microfluidics; zinc oxide; rhodamine B; rhodamine 6G

---

1. Introduction

With the advent of microfluidics and the development of many lab on a chip (LOC) applications, the need to ensure a good control and monitoring of temperature at the microscale with a high spatial and temporal accuracy started to merge. For instance, in biology and medicine, droplet microfluidics based on polymerase chain reaction, or digital polymerase chain reaction (dPCR), proved to be a highly sensitive technique for gene detection and gene sequencing [1,2]. This powerful technique relies on three major steps, which are performed at three different temperatures (95 $^\circ$C, 54 $^\circ$C and 72 $^\circ$C) and repeated for 30 or 40 cycles to amplify a diluted template deoxyribonucleic acid (DNA) sample enabling to reach a detectable amount of suitable fluorogenic probes. Nevertheless, despite some progress [3], the realization of an integrated microfluidic digital PCR microchip, where the whole PCR thermo-cycling process can be realized, is still hampered by the difficulty to measure and to control precisely the temperature inside the microfluidic channels.

During the last two decades, several methods, mainly based on micro-electro-mechanical systems (MEMS) technology, have been developed to ensure a good thermal control within micro-systems and micro-devices [4]. Nevertheless, most of these methods suffer from several limitations such as high cost, complicated manufacturing and poor spatial resolution. Laser induced fluorescence (LIF) technique has proven to circumvent many of these limitations [3–14]. Indeed, many fluorescent dyes exhibit a fluorescence intensity which depends strongly on the temperature. Therefore, such dyes may be used as molecular probes to measure temperature in a highly sensitive and very localized manner, both in space and time. Among the most efficient molecular probes for temperature measurements, rhodamine B (RhB) is the most widely used.

At low concentrations, the fluorescence intensity I_f (per volume unit) of a fluorescent dye depends generally on several parameters according to the following linear relation, which may be derived simply from Beer-Lambert law [15–17]:

$$I_f \propto \Phi I_0 \epsilon C \tag{1}$$

where Φ, I_0, ϵ and C are the fluorescence quantum yield, incident illumination intensity, molar absorptivity and dye concentration, respectively. For most temperature sensitive dyes, likewise RhB, it is generally observed that an increase in temperature results in a decrease of the fluorescence quantum yield Φ and the fluorescence life time. This is because the non-radiative processes related to thermal agitation (collisions with solvent molecules, intra-molecular vibrations and rotations) are more efficient at high temperatures [15,18]. One may cite nevertheless, the particular cases of rhodamine 6G (Rh6G) whose fluorescence increases linearly with temperature [6] and rhodamine 110 (Rh110) whose fluorescence does not depend on temperature. For the latter compounds, the increase or stability of fluorescence intensity versus temperature may be attributed to the formation of complexes (dimers, trimers, …). Such complexes are better shielded from inter-molecular collisions and have higher quantum yields leading hence to an increase or a stability of fluorescence intensity versus temperature [18]. As can be seen from the above equation (Equation (1)), factors such as a non-uniform illumination (I_0) or a non-uniform dye concentration C can lead to major errors and uncertainties on temperature measurement.

While uncertainty caused by a fluctuation of the laser illumination can be eliminated relatively easily by normalizing fluorescence intensity using a reference temperature-insensitive dye, such as Rh110, those which are caused by a fluctuation of the dye concentration in the microfluidic device are much more difficult to handle. Indeed, several processes can be at the origin of concentration fluctuations, such as adsorption of dye molecules on the channel walls [7], trapping of molecules inside the porous substrate of the microfluidic device (e.g., PDMS) or a non homogeneous distribution of the flow advected dye molecules along the microchannel, caused by the well-known Taylor dispersion phenomena [19,20]. The existence of all these processes may result in a local depletion or enrichment of the dye in the microfluidic device.

One should emphasize also another potential and important source of uncertainty on the measure of temperature in microfluidic systems, that is the heat dissipation within a tiny volume illuminated by a focused laser beam. To our knowledge, such an effect has been often neglected in the literature. In order to minimize the increase of temperature caused by the heating of the incident laser beam, one should either use a highly sensitive detection setup enabling to minimize the illumination power and/or increase the efficiency of the heat dissipation in the illuminated area of the sample by using a suitable material with a high thermal conductivity.

The main objective of this work is to give a proof of concept of a new approach for measuring the temperature inside microfluidic channels at high throughput and in a reliable manner using laser induced fluorescence (LIF) technique of rhodamine B dye confined in highly monodisperse zinc oxide microspheres. Indeed, besides its high chemical stability and photo-stability, zinc oxide exhibits a high thermal conductivity of about 100 to 150 W/m·K (depending on ZnO structural properties) [21–23]. This value is for instance two orders of magnitude higher than the thermal conductivity of water, i.e., 0.65 W/m·K. We used three types of dye doped ZnO microparticles: one type is doped with RhB, a second type is doped with Rh6G and the third type is doped with a mixture of RhB and Rh6G. The later mixture is intended to be used as a reference for fluorescence intensity calibration and to take account of fluctuations in incident illumination intensity. Indeed, the fluorescence intensity of RhB decreases linearly with increasing temperature at a ratio of about −2%/°C whereas the fluorescence intensity of rhodamine 6G (Rh6G) increases linearly with increasing temperature at a rate of about +2%/°C [6,12,24,25], as shown (see Figure 1).

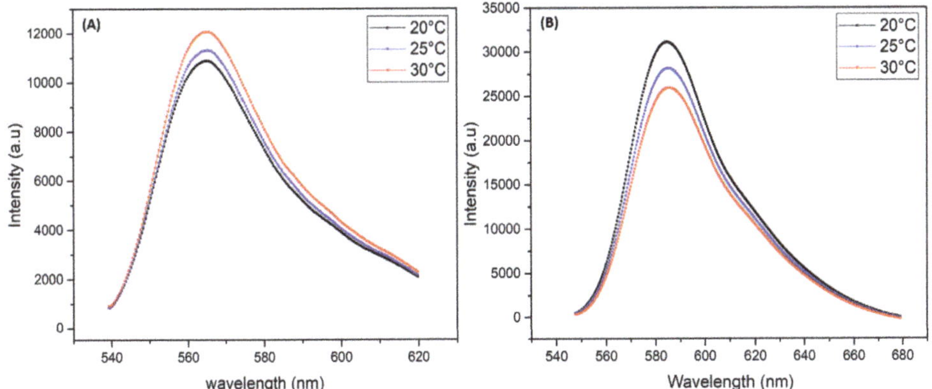

Figure 1. Fluorescence spectra of Rhodamine 6G (**A**) and Rhodamine B (**B**) versus temperature.

2. Experimental Section

2.1. Synthesis of ZnO Nanoparticles Building Units

We first synthesized ZnO nanoparticles which serve as building blocks for the final ZnO microparticles using sol-gel technique. Then, we confined such nanoparticles in highly monodisperse droplets using a flow-focusing microfluidic device in order to fabricate highly monodisperse ZnO microspheres where RhB and Rh6G are confined in a controlled manner.

The synthesis of colloidal ZnO nanoparticles can be carried out using different types of precursors dissolved in alcohol solvents [26]. We used zinc acetate dehydrate Zn(CH$_3$COO)$_2$:2H$_2$O (99.999%, from Sigma-Aldrich France, St. Louis, MO, USA) as ZnO precursor and methanol as a solvent, following the procedure which is described in more detail in [27]. All reagents were of analytical grade and were used as received without any further purification. Briefly, we dissolved 0.6 g in 5 mL of methanol (CH$_3$OH), then, the solution was stirred during 1 h at 60 °C under magnetic stirring to ensure homogeneous mixing and obtain a transparent solution.

The mechanism of formation of zinc oxide nanoparticles used in this work is according to a succession of chemical reactions. First, the dissolution of dehydrated zinc acetate in the presence of methanol contributes to the dehydration of zinc oxide precursor which results in the formation of anhydrous zinc acetate and water. Followed by a preliminary dissolution of the anhydrous zinc acetate to form the zinc ion and acetate. The latter will become acetic acid. It then takes place a chemical reaction between the species present (zinc and hydroxide ions) in the solution causing the precipitation of zinc hydroxide, which is ultimately converted to ZnO nanoparticles. Finally, the resulting dispersion of ZnO nanoparticles (with residual solvents) is doped with the desired fluorescent dye/dyes. Subsequently, the resulting mixture was immediately injected into the microfluidic chip to generate highly monodisperse microdroplets. The formation mechanism of zinc oxide nanoparticles can be summarized in these chemical reactions:

$$Zn(CH_3COO)_2 \cdot 2H_2O \longrightarrow Zn(CH_3COO)_2 + 2H_2O$$

$$Zn(CH_3COO)^{2-} \longrightarrow Zn^{2+} + 2CH_3COO^-$$

$$Zn^{2+} + 2OH^- \longrightarrow Zn(OH)_2$$

$$Zn(OH)_2 \longrightarrow ZnO + H_2O$$

2.2. Fabrication of Microfluidic Devices and Synthesis of ZnO Microparticles

Microfluidic devices were manufactured according to the conventional soft lithography technique [28]. In a first step, a pattern was transferred to SU-8 photoresist, previously coated on a silicon wafer, followed by the exposure to ultraviolet (UV) light, through the mask pattern. UV illumination leads to the polymerization of the photoresist located under the transparent regions of the mask. After development, the master mold is ready for the next step. In a second step, the polydimethylsiloxane (PDMS) was first mixed with a cross-linking agent with a weight ratio of 10:1, the mixture then was degassed using a vacuum pump at room temperature and the solution was poured onto the previously fabricated mold and placed in the oven for polymerization at 75 °C for 2 h. The block of PDMS was then removed from the mold; we thus obtain a replica of microchannels. In a third step, the PDMS block and the glass slide were treated with oxygen plasma for 20 s to enable their bonding and sealing the microfluidic chip. The design of the microfluidic device for droplets generation contained two inputs, one input for the carrier oil and a second input for the dispersed phase, a main (square) channel cross section of about 80 μm × 80 μm and an output for the collection of droplets in a Petri dish (see Figure 2a,b). The flow rates of the carrier oil (Q_c) and the dispersed phase (ZnO dispersion) (Q_d) were set using Nemesys syringe pumps (Cetoni GmbH, Korbussen, Germany). The dispersed phase consisted of ZnO solution doped with one of the selected rhodamine dyes: RhB, Rh6G and a mixture of both dyes. The flow rates were set in order to produce droplets with the same size (55 μm) for all experiments. Then, the droplets were collected in Petri-dish with HFE 7500 at room temperature.

The used carrier oil phase consisted of a fluorocarbon oil (HFE 7500, 3-ethoxy-dodecafluoro-2-trifluoromethyl-hexane, Inventec, Bry-sur-Marne, France), with a density of 1.61 g/cm^3 and thermal diffusivity of about $\kappa = 3.6 \times 10^{-8}$ m^2/s; 0.2 % (w/w) of a commercial surfactant (dSURF, Fluigent, Le Kremlin-Bicêtre, France) was added in HFE 7500 oil in order to prevent droplets merging. The used fluorocarbon oil has the advantage of not inducing PDMS swelling and being also chemically inert. More important for our application, it does not solubilize any of the ZnO precursor solution components.

After their generation, droplets were transported along the microfluidic channel by the carrier oil phase (see Figure 2) and then collected in a Petri dish (not shown on Figure 2), where droplets formed a floating monolayer at the oil-air interface, because of the higher density of fluorocarbon oil ($d_{HFE\text{-}7500} = 1.62$). In our study, the condensation of ZnO droplets was mainly controlled by the evaporation of the solvent at the oil–air interface. We let the condensation performing until microparticles reach their final sizes: 17.8 ± 0.2 μm for RhB doped microparticles and 20 ± 0.2 μm for both Rh6G and RhB/Rh6G doped microparticles.

For the re-injection of the synthesized microparticles, we used another microfluidic device (re-injector), which was made of a main inlet enabling the injection of the dispersed ZnO microparticles in HFE 7500 oil and three other secondary inlets for the injection of pure carrier oil to prevent aggregation of microparticles and the clogging of the microfluidic channel. The design of the re-injector includes also a constriction of the rectangular channel, with a width $w = 22$ μm and a height $h = 38$ μm.

The overall flow rate in this constriction area was about $Q = 150$ µL/h. Therefore a mean flow velocity of about $v_{constr.} = \dfrac{Q}{w \times h} \simeq 5$ cm/s can be deduced.

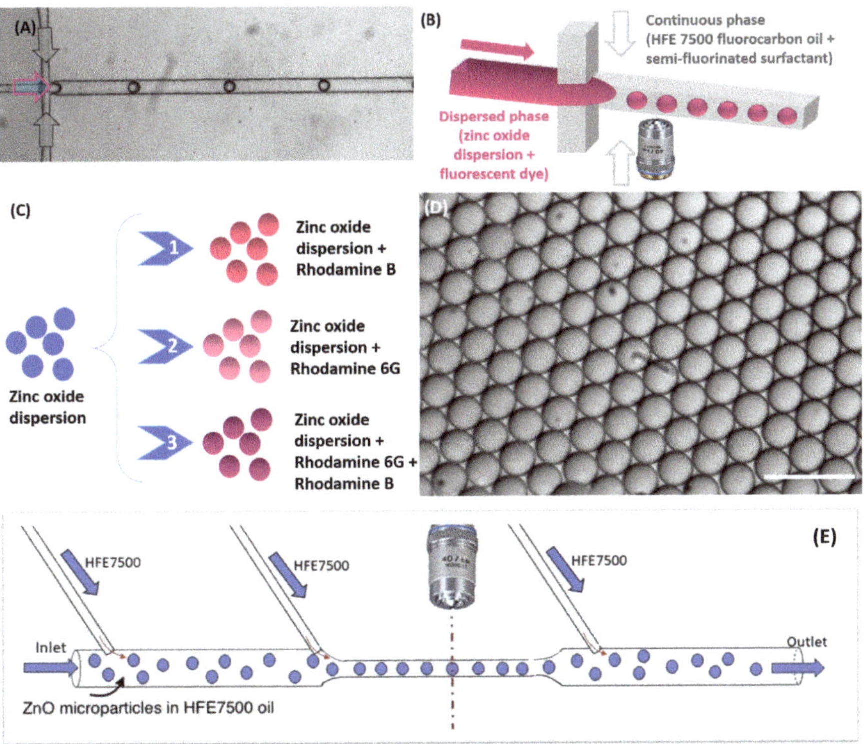

Figure 2. (**A**) Optical micrograph and schematic illustration of (**B**) the flow-focusing microfluidic geometry for zinc oxide droplet formation; (**C**) and the different case studies; (**D**) Optical micrograph of stable and monodisperse zinc oxide droplets generated through the flow-focusing microfluidic device, Scale bar, 100 µm; (**E**) Schematic illustration of the used microfluidic design for fluorescence analysis of doped ZnO microparticles.

2.3. Fluorescence Detection of Flowing Microparticles

Figure 3 shows the home-built fluorescence acquisition setup used in this study. It enables for a highly sensitive fluorescence detection, which is described in detail in Ref. [29]. It includes multiple laser sources optimised for the absorption of different fluorophores (we used in this study one continuous-wave (CW) source with wavelength of 532 nm). Laser incident beams are combined by mean of a first dichroic mirror (DM1) and then directed towards microdroplets in the microfluidic channel by mean of a second dichroic mirror (DM2). The focused band limited light is targeted towards the droplets/particles and recollected by the microscope objective, which is then transmitted through another set of band-limited filters to two photo-multiplier tubes (PMT's). The signal outputs from PMT's are then collected at high acquisition rates (100 KHz) using a DAQ acquisition card (National Instruments) and analyzed them using LabVIEW and a FPGA (field programmable gate arrays) module scripts, which allows for the identification of droplets by the modulation of fluorescence versus time. Depending on the selected sampling acquisition of the fluorescence signal, the width of the time lapse for signal detection can be varied from 40 ms (max sample rate of 200 KHz) to seconds or minutes.

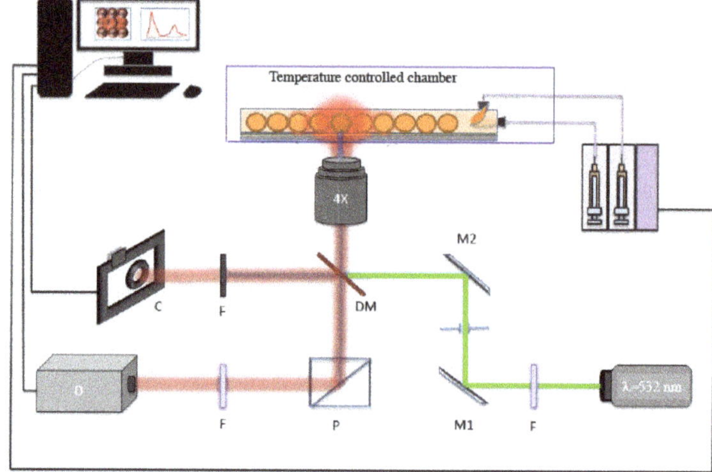

Figure 3. Experimental setup for high throughput fluorescence measurements.

2.4. Calibration of Temperature Measurements

We used a thermo-plate heating chamber (Tokai-Hit, Shizuoka, Japan) to control precisely the temperature of the whole microfluidic device (± 0.1 °C). This device is equipped with two heating plates, one at the bottom and one at the top. The later consists of a large clear glass top heater providing a uniform temperature distribution in the whole chamber. The heating device is also equipped with a feedback sensor mechanism that enables a real-time, precise sample temperature feedback temperature regulation. The heating chamber was placed on IX73 Olympus microscope (OLYMPUS, Tokyo, Japan) stage to measure the fluorescence of ZnO particles doped with rhodamine at different temperatures from 20 to 50 °C.

To determine the dependence of the fluorescence intensity versus temperature, we enclosed the entire microfluidic chip in the heating chamber. For each temperature value, we waited at least 10 min before recording the fluorescence intensity. Also, in order to give enough time for the incoming microparticles and HFE 7500 oil to reach the selected (target) temperature, a length of the tubing inlets (ZnO microparticles dispersion and pure fluorocarbon oil) of about $L \simeq 10$ cm was enclosed in the heating chamber. Taking account of the used overall flow rate, $Q \simeq 150$ µL/h and the inner diameter of the tubing, $D_{tub.} = 0.56$ mm, one may deduce a flow velocity in the tubing of about

$$v_{flow} = \frac{Q}{\pi (D_{tub.}/2)^2} \simeq 4 \text{ cm/min.}$$

This means that it should take less than 3 min for the incoming fluid and microparticles to flow from the entry of the heating system until the inlet of the microfluidic channel, where fluorescence intensity is measured.

Considering now the thermal diffusivity of HFE 7500 oil, that is 2.16 mm^2/min, and the cross section size of the tubing, one may calculate a diffusion time as small as $t_{diff} \simeq 4$ s, which is the time needed for the temperature to diffuse and become homogenized through all the cross section of the flowing fluid. As one may notice, the calculated diffusion time is much shorter than the flow time of the fluid before it reaches the area of the microfluidic channel where the temperature is measured.

3. Results and Discussion

3.1. Fluorescence Intensity Versus Temperature of ZnO Microparticles Doped with rhodamine B (RhB)

Figure 4 shows the change versus different temperatures, ranging from 20 to 50 °C, of the fluorescence intensity of RhB doped ZnO microparticles carried by a flow of HFE 7500 of about 150 µL/h in the microfluidic channel.

We first show in Figure 4A the fluorescence intensity peaks of individual flowing microparticles as they were detected and recorded from different experiments (at different temperatures). Because microparticles were randomly dispersed in the carrier oil, the time at which microparticles crossed the constriction area of the microfluidic channel, where they were detected, was random and hence the origin of the time axis was arbitrary. Nevertheless, one could have the measured value of the width of each peak to roughly estimate the size of the microspheres, which was found here to be about 1.5 ms, as shown in Figure 4B.

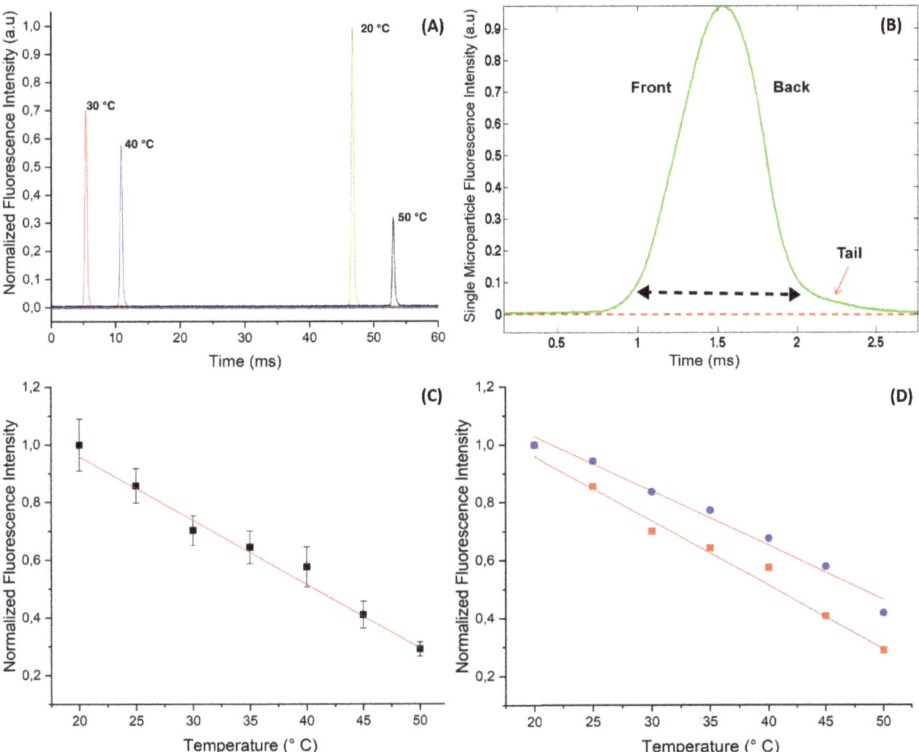

Figure 4. (**A**) Recorded fluorescence intensity peaks at different temperature from confined rhodamine B dye in flowing ZnO microcapsules (with a size of 17.8 µm) along a 22 µm wide micro-channel constriction area; (**B**) High-magnification of the fluorescence intensity peak vs. time of doped ZnO microcapsules with rhodamine B; (**C**) Plot of the linear decrease of the normalized fluorescence intensity of doped ZnO microparticles with rhodamine B vs. temperature; (**D**) Comparison of the change of fluorescence of RhB vs. temperature in flowing RhB doped ZnO microparticles (squares, red curve) and flowing RhB doped ZnO microdroplets (circles, blue curve), the slope of the two curves are very close and in very good agreement with the literature data: $(-2.2 \pm 0.1)\%/°C$ and $(-1.9 \pm 0.2)\%/°C$, respectively.

Taking account the mean flow velocity of 5 cm/s in the constriction area of the microchannel, one deduces a microparticle size of about $L \simeq 75$ μm, which was four times the size we measured from microscopy image analysis, $D_{RhB} = 17.8 \pm 0.2$ μm. Noteworthy, we made the assumption that microparticles flow with the same mean velocity as the surrounding carrier fluid's, which was basically not a good approximation, because of the presence of a thin lubrication film of the continuous phase between the particles and the channel walls as reported in the literature [30,31].

It is also interesting to notice from Figure 4B that the fluorescence peaks exhibited an asymmetric bell shape with an elongated tail at the back side of the microparticle. This observation, with other results obtained from optical and scanning electronic microscopies (not shown in this study), corroborates the relative softness of the synthesized microparticle and their deformability. We have shown in a previous study that the synthesized microparticles consist of microcapsules with a hollow structure and a thin envelope whose thickness value was found to be about 0.7 μm [27]. The group of Slasac and Barhès-Biesel et al. have shown for instance that microcapsules can deform easily under the effect of a shear flow in a microchannel [32,33].

One of the main advantage of droplet microfluidics based synthesis approach is the ability to obtain highly monodisperse microparticles, which in turn enabled us to obtain results with very good statistics. We normalized the fluorescence intensity values by considering the intensity recorded at 20 °C as a reference, $I_{20°C} = 1$.

For each temperature, we collected the fluorescence signal from approximately 20 highly monodisperse microparticles and plot the average values with their standard deviation versus temperature, as shown in Figure 4C. This figure shows clearly that the fluorescence intensity of RhB doped ZnO microparticles decreased linearly vs. temperature with a rate of about $(-2.2 \pm 0.1)\%/°C$. The determined slope was in very good agreement with the value reported generally in the literature for RhB dye [4–6,12–14].

In order to compare the effect of ZnO on the fluorescence behavior of RhB versus temperature, we present in Figure 4D the (normalized) fluorescence intensity of RhB when confined in microdroplets made of a water based RhB solution (0.5 mM and droplets size about 50 μm). We found that the slope of the linear decrease of the fluorescence intensity versus temperature was very close to the one recorded from flowing RhB doped ZnO microparticles, $(-1.9 \pm 0.2)\%/°C$.

3.2. Fluorescence Intensity Versus Temperature of ZnO Microparticles Doped with rhodamine 6G (Rh6G)

Figure 5A shows the change versus temperature of the fluorescence intensity of ZnO microparticles when doped with Rh6G dye. As may be noticed, though the fluorescence intensity appeared to increase globally with temperature, as observed in the bulk solution [6], we observed for the first time that the fluorescence intensity exhibited a minimum around 40 °C. This behavior is reproducible and was observed only when ZnO microparticles were doped with Rh6G.

Figure 5B shows a "standard" linear increase versus temperature of the of the fluorescence intensity of Rh6G when dissolved in microdroplets made of a 0.5 mM aqueous solution (droplet size equal to 50 μm) and when dissolved in a bulk solution. Nevertheless, the obtained slope for both curves, around 0.5%/°C, was much smaller than the one reported in the literature, 1.9%/°C [6]. Nevertheless, no minimum of intensity was observed in these cases, which indicates that this phenomena must be related to the confinement of Rh6G in ZnO microparticles.

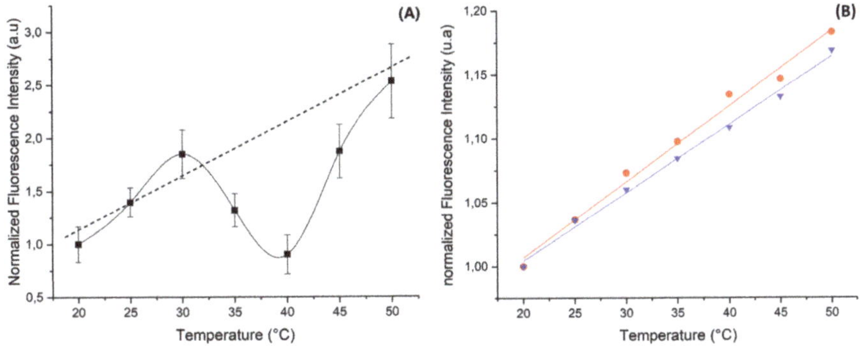

Figure 5. Change of the relative intensity of the fluorescence of rhodamine 6G (Rh6G) vs. temperature in (**A**) ZnO microparticles and (**B**) in water droplet (blue curve) and in ethanol droplet (red curve).

Though, the aim of this study is to give a simple demonstration of the proof of concept of our dye doped ZnO microparticles approach for locally measuring the temperature, we present preliminary results obtained by optical and scanning electron microscopies, which enables to understand (partly) the observed difference between RhB and Rh6G dyes in ZnO microparticles, as shown in Figure 6.

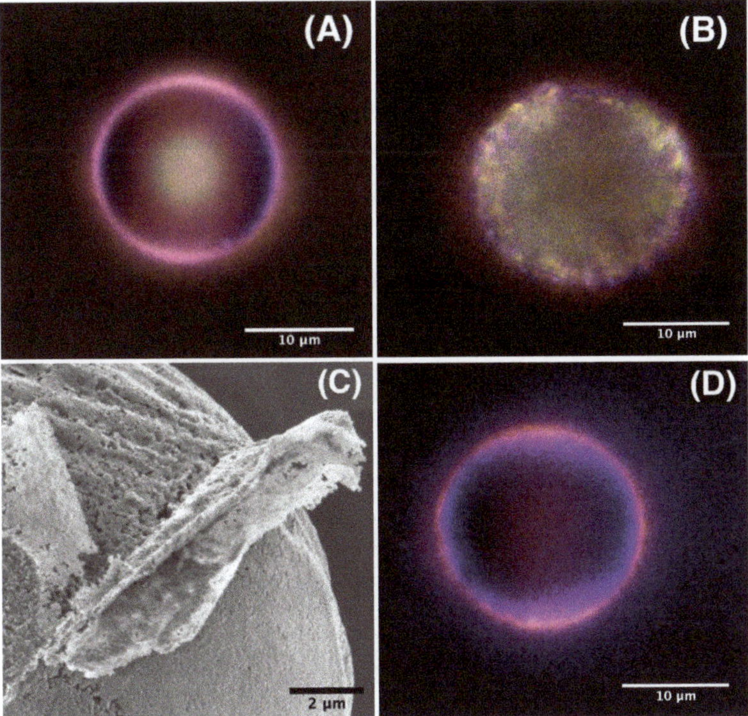

Figure 6. Optical microscopy images of doped ZnO microspheres with RhB (**A**), Rh6G (**B**) and a RhB/Rh6G mixture (**D**); (**C**) High resolution scanning electron microscopy image of a ZnO microparticle showing the aggregation of ZnO nanocrystal building units of the microsphere and the resulting porosity of the material.

Figure 6A shows that RhB molecules organized in a homogeneously distributed pink colored shell around ZnO microsphere, which corresponded basically, as stated earlier, to a microcapsule with a thin porous envelope having a thickness of about 0.7 µm, as shown in more detail in reference [27]. We emphasize in this study the porosity of the microcapsule shell, which resulted from the aggregation of ZnO nanoparticles building units, whose average size was found to be about 100 nm, and the formation of voids between such nanoparticles, as can be seen from the scanning electron microscope (SEM) image shown on Figure 6C. One may suggest also that these pores may serve as niches for aggregates of dyes molecules.

Also, since rhodamine molecules were added at the early stage of the formation of the microcapsules, i.e., in the microdroplet state, the observed colored ring on the surface of ZnO microsphere should correspond to an adsorbed layer of RhB molecules on the inner surface of the microcapsule. In contrast, Figure 6B shows that Rh6G dye molecules were less localized on the surface of the microcapsule and the dye seemed to occupy a diffuse and extended area all over the surface of the shell. Moreover, we observe the presence of many bright spots with different brightness, which revealed a non homogeneous distribution of Rh6G dye molecules inside the shell of the microcapsule. We attribute such brighter regions to pores and niches in the ZnO microcapsule shell where the dye accumulates more or less strongly. Moreover, though we have used the same size for the initial droplets (55 µm) and the same ZnO precursor concentration (0.25 mM) for both types of microcapsules, we found that those doped with Rh6G had a lager size (20 ± 0.2 µm) than those doped with RhB (17.8 ± 0.2 µm).

We suggest that the two different behaviors of the two dyes with ZnO microcapsules should be related to the electrical charges carried by the dyes and the electrically charged ZnO nanocrystals building blocks. Indeed, whereas RhB molecules can adopt (depending on the pH value) either a positively charged form or a zwitterionic form, for which the overall electrical charge is zero, Rh6G molecules can carry only a positive electric charge, as shown from their chemical structures in Figure 7A. Also, ZnO nanocrystals are known to carry a net electric dipole moment, which is inherent to the tetrahedral configuration of the wurtzite crystalline structure. In a such structure, each type of ion, Zn^{2+} or O^{2-}, has four neighbouring ions of the other type of atom, and vice versa. That is to say, the tetrahedral coordination exhibits a sequence of positively charged Zn^{2+} and negatively charged O^{2-} polar planes within the c-axis direction, thus respectively contributing to two opposite faces of polarity (0001) and $(000\bar{1})$ perpendicular to the c-axis. This results in an intrinsic electric dipole moment and a spontaneous polarization along the c-axis of the nanocrystals [34–38].

Therefore, one may expect that a strong interaction should occur between the positively charged Rh6G molecules and the negatively charged plane of ZnO nanocrystals, leading probably to a larger diffusion of Rh6G inside the porous structure of the microcapsule shell than could occur with RhB molecules in their zwitterionic form. At this stage, we still need further investigations in order to make a correlation between the adsorption of Rh6G dye in the pores of the shell and the observation of a minimum fluorescence intensity at a temperature around 40 °C.

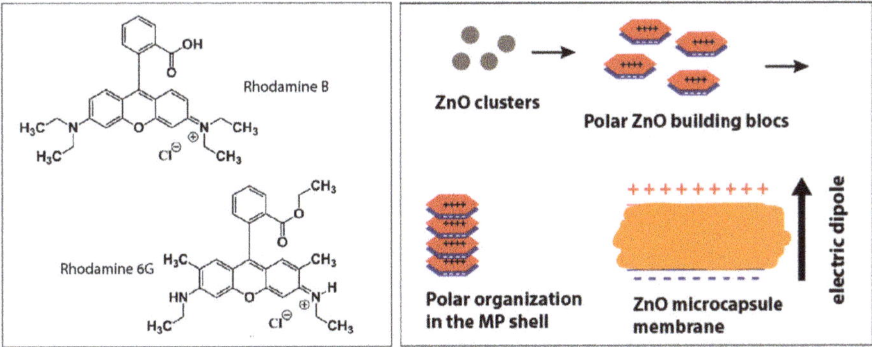

Figure 7. (**left**) Chemical structure of the used RhB and Rh6G dyes, depending on the pH, RhB molecules may exhibit either a zwitterionic form (presence of both a positive and negative charges are present) or may exhibit only a negative charge, whereas Rh6G molecules can carry only a positive charge or no charge at all, (**right**) illustration of the formation of the polar ZnO nanocrystals building units.

3.3. Fluorescence Properties of ZnO Microparticles Doped with a Mixture of RhB and Rh6G

Figure 8 shows the fluorescence intensity versus temperature of ZnO microparticles which were doped with a mixture of RhB (29 µm) and Rh6G (37 µm) dyes. Because of the non linear variation of the fluorescence intensity of Rh6G in ZnO microparticles, it was difficult to predict the concentration ratio of the two dyes which would lead to a constant fluorescence intensity versus temperature. Nevertheless, after several attempts, we found that the ratio $\frac{[RhB]}{[Rh6G]} = \frac{29\ \mu M}{37\ \mu M}$ gave a good reference for which the fluorescence intensity remained practically constant versus temperature, as shown in Figure 8.

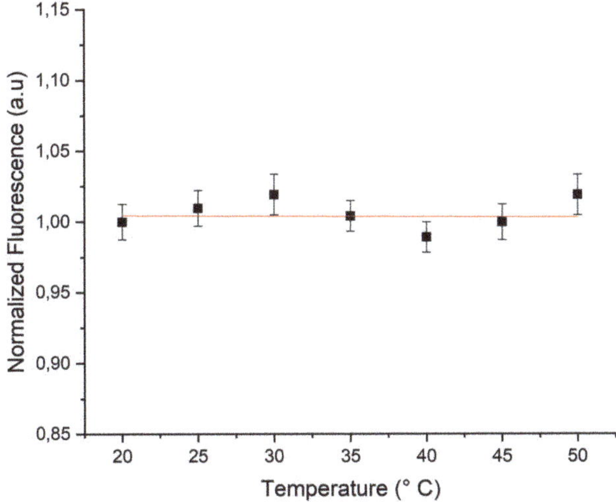

Figure 8. Relative intensity of the fluorescence of RhB and Rh6G mixture versus temperature in ZnO microparticles.

4. Conclusions and Perspectives

We have shown in the present study that ZnO microparticles doped with RhB dye exhibit a fluorescence intensity which decreases linearly versus temperature with a rate of about $(-2.2 \pm 0.1)\%/°C$. This value is in very good agreement with the literature date. Our results should pave the way for the development of a promising technique which will enable in the future for a real time and highly sensitive measurement of the temperature in microfluidic channels. The use of ZnO microparticles for confining the temperature sensitive dyes has two main advantages: (i) it enables to circumvent the diffusion and the adsorption of the dyes on the walls of the microfluidic channel and inside the pores of the PDMS substrate of the microfluidic device, (ii) it enhances the dissipation of the heat generated by the focused incident laser beam thanks to the high thermal conductivity of ZnO material. In contrast with the fluorescence intensity linear change versus temperature of Rh6G in bulk solution or in microdroplets, Rh6G when confined in ZnO microparticles shows a non linear behavior of the fluorescence intensity versus temperature. The presence of a minimum at a temperature around 40 °C is reported for the first time. Such non linear behavior makes the construction of reliable reference a little bit cumbersome by nonetheless we determined experimentally a suitable mixture of RhB and Rh6G for which the fluorescence intensity remains practically constant in the investigated temperature range [20 °C–50 °C]. We envisage to consider in a future study the use of rhodamine 110 for the construction of such reference. Also, we plan to flow and detect simultaneously the two types of microparticles (RhB as a measure probe and Rh110 as reference probe) in the microchannel in order to make the normalization of the fluorescence intensity signal straightforward. For this purpose, the use of two different concentrations for the two types of microparticles will allow for the discrimination between the two types of particles while flowing at high throughput in the microchannel, as illustrated in Figure 9.

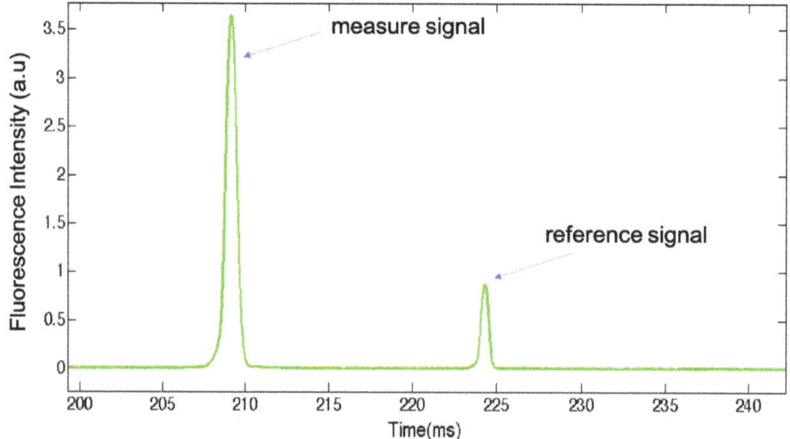

Figure 9. Different microparticles doped with two different dyes may be discriminated by the value of their fluorescence intensity signals. Such concentration based coding method will be applied in the future for the simultaneous recording of the fluorescence intensity signals from RhB doped microparticles (measure probes) and Rh110 doped microparticles (reference probes).

Author Contributions: Conceptualization, A.I.E.A.; methodology, N.G., S.R. and A.T.; validation, A.I.E.A. and A.C.; formal analysis, N.G., S.R. and A.I.E.A.; investigation, N.G. and S.R.; resources, Z.H., A.C. and A.I.E.A.; writing—original draft preparation, N.G., S.R. and A.I.E.A.; writing–review and editing, N.G., Z.H., A.T., A.C. and A.I.E.A.; visualization, Z.H.; supervision, A.C. and A.I.E.A.; project administration, A.C. and A.I.E.A.; funding acquisition, A.C. and A.I.E.A. All authors have read and agreed to the published version of the manuscript.

Funding: This research was funded by PHC-Toubkal 2017 program (Programme Hubert Curien).

Conflicts of Interest: The authors declare no conflict of interest.

References

1. Pohl, G.; Shih, I.M. Principle and applications of digital PCR. *Expert Rev. Mol. Diagn.* **2004**, *4*, 41–47. [CrossRef]
2. Taly, V.; Pekin, D.; El Abed, A.; Laurent-Puig, P. Detecting biomarkers with microdroplet technology. *Trends Mol. Med.* **2012**, *18*, 405–416. [CrossRef]
3. Guijt, R.M.; Dodge, A.; van Dedem, G.W.; de Rooij, N.F.; Verpoorte, E. Chemical and physical processes for integrated temperature control in microfluidic devices. *Lab Chip* **2003**, *3*, 1–4. [CrossRef]
4. Gosse, C.; Bergaud, C.; Löw, P. Molecular probes for thermometry in microfluidic devices. *Thermal Nanosystems and Nanomaterials*; Springer: Berlin, Germany 2009; pp. 301–341.
5. Ross, D.; Gaitan, M.; Locascio, L.E. Temperature measurement in microfluidic systems using a temperature-dependent fluorescent dye. *Anal. Chem.* **2001**, *73*, 4117–4123. [CrossRef] [PubMed]
6. Kuzkova, N.; Popenko, O.; Yakunov, A. Application of temperature-dependent fluorescent dyes to the measurement of millimeter wave absorption in water applied to biomedical experiments. *J. Biomed. Imaging* **2014**, *2014*, 12. [CrossRef] [PubMed]
7. Samy, R.; Glawdel, T.; Ren, C.L. Method for microfluidic whole-chip temperature measurement using thin-film poly (dimethylsiloxane)/Rhodamine B. *Anal. Chem.* **2008**, *80*, 369–375. [CrossRef]
8. Kim, M.M.; Giry, A.; Mastiani, M.; Rodrigues, G.O.; Reis, A.; Mandin, P. Microscale thermometry: A review. *Microelectron. Eng.* **2015**, *148*, 129–142. [CrossRef]
9. Christofferson, J.; Maize, K.; Ezzahri, Y.; Shabani, J.; Wang, X.; Shakouri, A. Microscale and nanoscale thermal characterization techniques. *J. Electron. Packag.* **2008**, *130*, 041101. [CrossRef]
10. Erickson, D.; Sinton, D.; Li, D. Joule heating and heat transfer in poly (dimethylsiloxane) microfluidic systems. *Lab Chip* **2003**, *3*, 141–149. [CrossRef] [PubMed]
11. Arata, H.F.; Löw, P.; Ishizuka, K.; Bergaud, C.; Kim, B.; Noji, H.; Fujita, H. Temperature distribution measurement on microfabricated thermodevice for single biomolecular observation using fluorescent dye. *Sens. Actuators B: Chem.* **2006**, *117*, 339–345. [CrossRef]
12. Liu, G.; Lu, H. Laser-induced fluorescence of rhodamine B in ethylene glycol solution. *Procedia Eng.* **2015**, *102*, 95–105. [CrossRef]
13. Ebert, S.; Travis, K.; Lincoln, B.; Guck, J. Fluorescence ratio thermometry in a microfluidic dual-beam laser trap. *Opt. Express* **2007**, *15*, 15493–15499. [CrossRef] [PubMed]
14. Sakakibara, J.; Adrian, R.J. Whole field measurement of temperature in water using two-color laser induced fluorescence. *Exp. Fluids* **1999**, *26*, 7–15. [CrossRef]
15. Valeur, B.; Berberan-Santos, M. Molecular Fluorescence: Principles and Applications. *J. Biomed. Opt.* **2013**, *18*, 039901.
16. Guilbault, G.G. *Practical Fluorescence, Theory, Methods and Techniques*; Marcell Dekker: New York, NY, USA, 1973.
17. Chamarthy, P.; Garimella, S.V.; Wereley, S.T. Measurement of the temperature non-uniformity in a microchannel heat sink using microscale laser-induced fluorescence. *Int. J. Heat Mass Transf.* **2010**, *53*, 3275–3283. [CrossRef]
18. Hossain, M.A.; Canning, J.; Yu, Z.; Ast, S.; Rutledge, P.J.; Wong, J.K.H.; Jamalipour, A.; Crossley, M.J. Time-resolved and temperature tuneable measurements of fluorescent intensity using a smartphone fluorimeter. *Analyst* **2017**, *142*, 1953–1961. [CrossRef]
19. Taylor, G. Dispersion of a solute in a solvent under laminar conditions. *Proc. R. Soc. London, Ser. A* **1953**, *219*, 186–203.
20. Beard, D.A. Taylor dispersion of a solute in a microfluidic channel. *J. Appl. Phys.* **2001**, *89*, 4667–4669. [CrossRef]
21. Özgür, Ü.; Gu, X.; Chevtchenko, S.; Spradlin, J.; Cho, S.J.; Morkoç, H.; Pollak, F.; Everitt, H.; Nemeth, B.; Nause, J. Thermal conductivity of bulk ZnO after different thermal treatments. *J. Electron. Mater.* **2006**, *35*, 550–555. [CrossRef]
22. Florescu, D.I.; Mourokh, L.; Pollak, F.H.; Look, D.C.; Cantwell, G.; Li, X. High spatial resolution thermal conductivity of bulk ZnO (0001). *J. Appl. Phys.* **2002**, *91*, 890–892. [CrossRef]
23. Alvarez-Quintana, J.; Martínez, E.; Pérez-Tijerina, E.; Pérez-García, S.; Rodríguez-Viejo, J. Temperature dependent thermal conductivity of polycrystalline ZnO films. *J. Appl. Phys.* **2010**, *107*, 063713. [CrossRef]

24. Zehentbauer, F.M.; Moretto, C.; Stephen, R.; Thevar, T.; Gilchrist, J.R.; Pokrajac, D.; Richard, K.L.; Kiefer, J. Fluorescence spectroscopy of Rhodamine 6G: Concentration and solvent effects. *Spectrochim. Acta A Mol. Biomol. Spectrosc.* **2014**, *121*, 147–151. [CrossRef] [PubMed]

25. Ali, M.; Moghaddasi, J.; Ahmed, S. Temperature effects in Rhodamine B dyes and improvement in CW dye laser performance. *Laser Chem.* **1991**, *11*, 31–38. [CrossRef]

26. Hamzaoui, N.; Boukhachem, A.; Ghamnia, M.; Fauquet, C. Investigation of some physical properties of ZnO nanofilms synthesized by micro-droplet technique. *Results Phys.* **2017**, *7*, 1950–1958. [CrossRef]

27. Ghifari, N.; Chahboun, A.; El Abed, A. One-Step Synthesis of Highly Monodisperse ZnO Core-Shell Microspheres in Microfluidic Devices. In Proceedings of the 2019 21st International Conference on Transparent Optical Networks (ICTON), Angers, France, 9–13 July 2019; pp. 1–6.

28. Duffy, D.C.; McDonald, J.C.; Schueller, O.J.; Whitesides, G.M. Rapid prototyping of microfluidic systems in poly (dimethylsiloxane). *Anal. Chem.* **1998**, *70*, 4974–4984. [CrossRef] [PubMed]

29. Hayat, Z.; El Abed, A.I. High-throughput optofluidic acquisition of microdroplets in microfluidic systems. *Micromachines* **2018**, *9*, 183. [CrossRef]

30. Taylor, G. Deposition of a viscous fluid on the wall of a tube. *J. Fluid Mech.* **1961**, *10*, 161–165. [CrossRef]

31. Baroud, C.N.; Gallaire, F.; Dangla, R. Dynamics of microfluidic droplets. *Lab Chip* **2010**, *10*, 2032–2045. [CrossRef]

32. Chu, T.X.; Salsac, A.V.; Barthès-Biesel, D.; Griscom, L.; Edwards-Lévy, F.; Leclerc, E. Fabrication and in situ characterization of microcapsules in a microfluidic system. *Microfluid. Nanofluidics* **2013**, *14*, 309–317. doi:10.1007/s10404-012-1049-9. [CrossRef]

33. Hu, X.Q.; Salsac, A.V.; Barthès-Biesel, D. Flow of a spherical capsule in a pore with circular or square cross-section. *J. Fluid Mech.* **2012**, *705*, 176–194. doi:10.1017/jfm.2011.462. [CrossRef]

34. Liu, Z.; Wen, X.; Wu, X.; Gao, Y.; Chen, H.; Zhu, J.; Chu, P. Intrinsic dipole-field-driven mesoscale crystallization of core- shell ZnO mesocrystal microspheres. *J. Am. Chem. Soc.* **2009**, *131*, 9405–9412. [CrossRef] [PubMed]

35. Tampo, H.; Fons, P.; Yamada, A.; Kim, K.K.; Shibata, H.; Matsubara, K.; Yoshikawa, H.; Kanie, H.; Niki, S. Determination of crystallographic polarity of ZnO bulk crystals and epilayers. *Phys. Status Solidi C* **2006**, *3*, 1018–1021. [CrossRef]

36. Dai, S.; Park, H.S. Surface effects on the piezoelectricity of ZnO nanowires. *J. Mech. Phys. Solids* **2013**, *61*, 385–397. [CrossRef]

37. Xiang, H.; Yang, J.; Hou, J.; Zhu, Q. Piezoelectricity in ZnO nanowires: a first-principles study. *Appl. Phys. Lett.* **2006**, *89*, 223111. [CrossRef]

38. Yufei, Z.; Zhiyou, G.; Xiaoqi, G.; Dongxing, C.; Yunxiao, D.; Hongtao, Z. First-principles of wurtzite ZnO (0001) and (0001) surface structures. *J. Semicond.* **2010**, *31*, 082001. [CrossRef]

 *micromachines*

Article

Synthesis and a Photo-Stability Study of Organic Dyes for Electro-Fluidic Display

Yong Deng [1,2,3], Shi Li [1], Dechao Ye [2,3], Hongwei Jiang [1], Biao Tang [1,*] and Guofu Zhou [1,2,3]

[1] Guangdong Provincial Key Laboratory of Optical Information Materials and Technology & Institute of Electronic Paper Displays, South China Academy of Advanced Optoelectronics, South China Normal University, Guangzhou 510006, China; yong.deng@guohua-oet.com (Y.D.); shili@m.scnu.edu.cn (S.L.); hongwei.jiang0822@gmail.com (H.J.); guofu.zhou@m.scnu.edu.cn (G.Z.)

[2] Shenzhen Guohua Optoelectronics Tech. Co. Ltd., Shenzhen 518110, China; dechao.ye@guohua-oet.com

[3] Academy of Shenzhen Guohua Optoelectronics, Shenzhen 518110, China

[*] Correspondence: tangbiao@scnu.edu.cn; Tel.: +020-39314813

Received: 9 December 2019; Accepted: 7 January 2020; Published: 11 January 2020

Abstract: Electro-fluidic display (EFD) is one of the most promising reflective displays because of its full color and video speed. Colored EFD oil, which normally consists of soluble organic dyes and non-polar solvent, plays a critical role in color, electro-optical behavior, and the reliability of the EFD devices. In this paper, we report our research on two kinds of electro-fluidic dyes based on anthraquinone and azo pyrazolone, including their synthesis, structure characterization, and application properties. Changes of absorbance curves, color coordinates of oils, and photoelectric responses of devices were studied in detail under accelerated irradiation to investigate the photo-stability and reliability properties of synthesized oil materials and devices. Photoelectric responses and photo stability of dyes are highly varied depending on their structures. We found that 1,4-dlialkylamino anthraqinone and mono azo pyrazolone dyes are much more stable than 1,8-dlialkylamino anthraqinone and corresponding bisazo pyrazolone dyes.

Keywords: electro-fluidic display; organic dye; colored oil; photo-stability

---

1. Introduction

Recently, electro-fluidic display as an emerging display technology has received widespread attention [1–4]. It has many advantages: (1) a reflective mode for using ambient light and energy saving [5]; (2) quick response (its switching time is less than 20 ms) for video display [6]; (3) superb optical performance (its white state reflectance is up to 50% [7] and it has full color) [8]; and (4) fluidic and soft display candidate materials for flexible displays.

Colored oils act as optical switches in electro-fluidic display, which affect not only the color gamut but also the contrast ratio, electro-optical response, and the lifetime for outdoor display devices. High optical density, good solubility in non-polar solvent, and good light stability are the main characteristics for the designation of soluble organic dyes. A series of efforts to develop efficient dyes for electro-fluidic display have been conducted in previous research, such as the designation of anthraquinone dyes [9,10], azo dyes [11–15], dipyrrole methane metal dyes [16], and pigment dispersion [17,18]. However, for many of the dyes that were disclosed in patents, few were subjected to detailed research on their synthesis and specific application properties, especially their photo-stability properties.

In our opinion, the photo-stability of the oil materials is one of color dyes' most crucial properties, which could limit their industrialization. In their designation lifetime, color dyes will decompose and their color image as well as their response properties will deteriorate. In this context, we have synthesized two typical electro-fluidic display (EFD) dyes (anthraquinone and azo pyrazolone dyes), and their electro-optical behavior, photo-stability properties, and structure correlation were researched in detail in this paper.

2. Experimental

2.1. Materials

All of the color dyes were synthesized in our laboratory. The structures were characterized and proved by using several analysis methods: Infrared (IR) (PerkinElmer, Shelton, CT, USA), Nuclear magnetic resonance (^{1}H NMR) (Bruker BioSpin GmbH, Rheinstetten, Germany), ^{13}C NMR (Bruker BioSpin GmbH, Rheinstetten, Germany), High resolution mass spectra (HRMS/ESI) (Bruker Daltonics, Bremen, Germany)). Indium Tin Oxide (ITO) coated glass (0.7 mm thickness) with a resistance of 100 $\Omega \cdot$m (purchased from Shenzhen Laibao Hi-Tech Co. Ltd., Shenzhen, China) was used as the bottom and top substrates. SU-8 3005 photoresist was purchased from Microchem Corp. (Newton, MA, USA). AF Teflon® 1600X was purchased from Dupont™ Co. (Wilmington, DE, USA)

2.2. Analysis Methods

UV-3300 spectrophotometer (Jinpeng Analytical Instrument Corporation, Shanghai, China) was used to record UV-vis (ultraviolet) absorption spectra of synthesized dyes. Perkin–Elmer 841 spectrometer (PerkinElmer, Shelton, CT, USA) using KBr pellets was used to record FTIR spectra. Varian AS400 (Agilent, Santa Clara, CA, USA), Bruker AVANCEIII 500 (Bruker BioSpin GmbH, Rheinstetten, Germany) and AVANCE NEO Bruker 600 (Bruker BioSpin GmbH) were used to record ^{1}H NMR and ^{13}C NMR spectra in deuterated solvents using tetramethylsilane (TMS) as an internal standard. Bruker Daltonics Microflex mass spectrometer (Bruker Daltonics, Bremen, Germany) was used to record the mass spectra. The thickness of the AF layer and height of the pixel wall were measured by using the Dektak (Dektak XT, BRUKER, Hamburg, Germany). A self-assembling oil filling instrument was made by our team. The photo-stability of Four dyes in EFD cell was tested in a Xenon arc lamp weather resistance test chamber (B-SUN-I), purchased from Shanghai Yiheng Instruments Co. Ltd (Shanghai, China). The simulation conditions were set according to international standard IEC 60068-2-5: Procedure B; 0.55 W/m^2 (340 nm); Temperature 45 °C. Absorbance curves before and after a period of irradiation were tested.

2.3. EFD Oil Formulation

Certain quantities of dyes and decane were weighted in a sample cell and put in the ultrasonic instrument for 30 min. Afterwards, the solution was subjected to a filtration process using a 0.45 μm filtrator to remove suspended impurities. Then the well prepared oil was used in the next EFD device fabrication process.

2.4. Electro-Fluidic Display (EFD) Device Fabrication

The EFD device was fabricated as follows: a 8 × 6 cm^2 ITO coated glass was used as a bottom substrate. A Teflon (AF Teflon® 1600X, Dupont™ Co.) layer of 0.8 μm was spin-coated on top of the ITO glass as the dielectric layer and hydrophobic layer. A pixel wall structure was fabricated by standard lithography process using photoresist material (SU-8 3005). After coating a photoresist film on active fluoropolymer surface, an exposure process was used to obtain a patterned reacted resist layer. Finally, a 7.5 μm height pixel wall with a 15 μm width was developed. The square of each pixel was 150 × 150 μm^2. After that, oil with an average thickness of 4.5 μm was filled into pixels by a self-assembly machine. The last step of the fabrication processes was to integrate another top ITO glass with a gap of 75 μm filled by a water phase.

3. Dye Synthesis and Characterization

The structure of four designed dyes were shown in the Figure 1.

Anthra-1: 2.0 g (242 g/mol, 0.0082 mol) Anthracene-1,4,9,10-tetraol was dissolved in 20 mL 2-methoxyethanol in a three-necked flask with agitation, and the solution was heated to 80 °C under an N_2 atomosphere. Then, 4.26 g (129 g/mol, 0.033 mol) 2-ethyl-hexylamine was added once. The solution was refluxed for 24 h. After that, the reaction was exposed to air for 2 h at 50 °C. Then the reaction was cooled to room temperature, poured into 100 mL water, and filtrated to get 2.77 g crude products Anthra-1. The crude Anthra-1 was purified by flash chromatography on silica gel (eluting agent: petroleum ether and ethyl acetate). Characterization: cyan solid; yield: 73.1%; λ (max) = 650 nm (in decane); ε = 17930 L·mol^{-1}·cm^{-1}. IR (KBr, cm^{-1}): 3061.80 (Ar–H); 2962.50 (–CH$_3$), 2921.20 (–CH$_2^-$), 2859.10 (–CH$_3$); 1645.9 (–C=O); 1581.80 (Ar); 1555.0 (Ar); 1521.90 (Ar); 1459.90 (–CH–); 1379.27(–CH$_3$); 1257.28 (C–N–C); 816.89. ^{1}H NMR (500 MHz, CDCl$_3$): 9.71 (s, 1H, Ar–NHCH$_2^-$), 8.25 (m, 2H, Ar–H), 7.65 (m, 2H, Ar–H), 7.35 (d, 2H, J = 19.0, Ar–H), 3.22 (m, 4H, –NCH$_2^-$), 1.73 (m, 2H, –CH–), 1.50 (m, 4H, –CH$_2^-$), 1.26 (m, 6H, –CH$_2^-$),1.18 (m, 6H, –CH$_2^-$), 0.88 (m, 12H, –CH$_3$). ^{13}C NMR (500 MHz, CDCl$_3$): 182.11, 146.56, 134.57, 131.84, 126.03, 123.66, 109.67, 45.99, 39.47, 31.28, 29.04, 24.53, 23.03, 14.12, 10.99. MS (MALDI-TOF) (DIF) M/Z (%): 463.20 (M + H)$^+$. Calc. for (C$_{30}$H$_{42}$N$_2$O$_2$): 462.30.

Anthra-1   **Anthra-2**

Pyrazolone-1   **Pyrazolone-2**

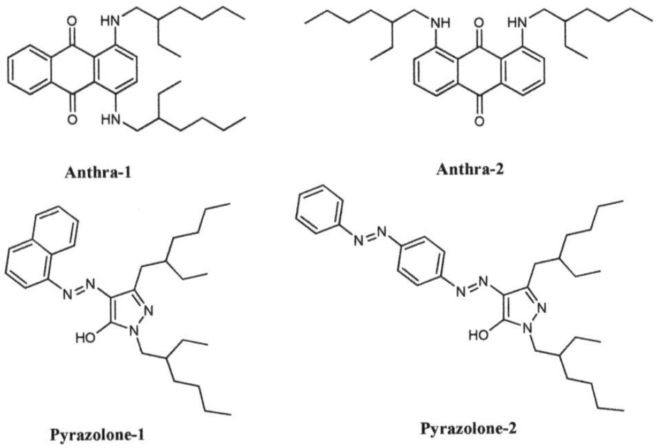

Figure 1. Molecular structures of the four synthesized dyes.

Anthra-2: 2.0 g (227 g/mol, 0.0088 mol) 1,8-dichloroquinone was dissolved in 20 mL 2-ethyl-hexylamine in a three-necked flask with agitation, and the solution was heated to 150 °C under an N_2 atomosphere for 12 h. After reaction, the solution was poured in 10% hydrochloric acid solution filtrate to remove excess 2-ethyl-hexylamine and to obtain 1.60 g of crude products Anthra-2. The crude Anthra-2 was purified by flash chromatography on silica gel (eluting agent: petroleum ether and ethyl acetate). Characterization: magenta solid; yield: 40%; λ (max) = 545 nm (in decane); ε = 13335 L·mol^{-1}·cm^{-1}. IR (KBr, cm^{-1}): 3268.50 (–NH–); 3082.40 (Ar–H); 2962.50 (–CH$_3$); 2925.30 (–CH$_2^-$); 2863.30 (–CH$_3$); 1658.30 (–C=O); 1623.20 (–C–N–); 1569.40 (Ar); 1531.60 (Ar); 1459.90 (–CH–); 1397.80 (–CH$_3$); 1296.50 (C–N–C); 1211.80 (C–N–C); 1073.20; 1019.51; 837.57; 744.53. ^{1}H NMR (500 MHz, CDCl$_3$): 9.67 (s, 1H, Ar–NH–); 7.45–7.43 (d, 2H, J =1 Hz, Ar–H); 7.40–7.37 (t, 2H, J = 15.5 Hz, Ar–H); 6.95–6.93 (d, 2H, J = 1 Hz, Ar–H); 3.16–3.14 (s, 4H, J = 11 Hz, Ar–NH–), 1.65 (m, 2H, –CH–); 1.44 (m, 8H, –CH2–); 1.28 (m, 8H, –CH2–); 0.91-0.88 (t, 6H, J =15 Hz, –CH$_3$); 0.86–0.83 (t, 6H, J = 15 Hz, –CH$_3$). ^{13}C NMR (500 MHz, CDCl$_3$): 188.99, 184.84, 151.48, 134.22, 117.63, 114.63, 46.08, 38.97, 31.49, 28.99, 24.79, 23.06, 14.11, 11.13. MS (MALDI-TOF) (DIF) M/Z (%): 463.20 (M + H)$^+$. Calc. for (C$_{30}$H$_{42}$N$_2$O$_2$): 462.30.

Pyrazolone-1: A quantity of 1.43 g (143 g/mol, 0.01 mol) naphthyl amine was dissolved in 20 mL of ethanol, 10 mL of water, and 3 mL (0.036 mol) of concentrated hydrochloric acid. After it had been rapidly cooled to 0–5 °C, a 5 mL solution containing 0.72 g (69 g/mol, 0.0105 mol) of sodium nitrite was added. Ehrlich reagent was used to detect the termination of diazotisation, and sulfamic acid was used to remove the residual nitrous acid. Afterwards, 2.94 g of the coupling component

1-(2-ethylhexyl)-3-(2-ethylhexyl)-5-pyrazalone (294 g/mol, 0.01 mol) was dissolved in 100 mL ethanol and cooled to 0–10 °C. Diazonium salt solution was added to the above solution at a temperature below 10 °C over a 15–20 min period, with the pH level maintained at 8–9 using a sodium hydroxyl solution. The reaction mixture was stirred for a further 3 h. Pyrazolone-1 was precipitated by adjusting the pH of the coupling solution to 4 using acetic acid, then collected and dried to get 3.5 g of crude product Pyrazolone-1. The crude Pyrazolone-1 was purified by flash chromatography on silica gel (eluting agent: ethyl acetate / petroleum ether = 1/100, v/v). Characterization: yellow solid; yield: 80.5%; λ (max) = 420 nm (in decane); ε =16119 L·mol^{-1}·cm^{-1}. IR (KBr, cm^{-1}): 3058.90 (Ar–H); 2958.60 (–CH$_3$); 2929.70 (–CH$_2$$^-$); 2856.40 (–CH$_3$); 1656.70 (C=O), 1560.40 (Ar); 1521.70 (Ar), 1456.10 (–N=N–); 1392.50 (–CH$_3$); 1244.00 (C–N–C); 1091.60, 792.70. ^{1}H NMR (500 MHz, CDCl$_3$): 14.478 (s, 1H, O–H); 8.00–7.99 (d, 1H, Ar–NH–N=Ar); 7.81–7.76 (d, 2H, Ar–H); 7.61–7.59 (d, 1H, Ar–H); 7.52-7.49 (t, 1H, Ar–H); 7.46–7.43 (m, 2H, Ar–H); 3.65–3.63 (d, 2H, –NCH$_2$$^-$); 2.66 (m, 1H,–CH–); 1.80 (m, 5H, –CH$_2$$^-$); 1.24 (m, 12H, –CH$_2$$^-$); 0.81–0.78 (m, 12H, –CH$_3$). ^{13}C NMR (500 MHz, CDCl$_3$): 159.03, 149.97, 136.36, 134.07, 129.88, 128.75, 127.00-123.52, 121.34, 110.76, 47.84, 38.47, 32.95, 30.65, 28.76, 26.22, 25.45-22.28, 14.11, 10.74. MS (MALDI-TOF) (DIF) M/Z (%): 461.30 (M–H)$^-$. Calc. for (C$_{29}$H$_{42}$N$_4$O): 462.30.

Pyrazolone-2: A quantity of 1.97 g (197 g/mol, 0.01 mol) 4-aminodiazobenzene was dissolved in 30 mL of water and 3 mL (0.036 mol) of concentrated hydrochloric acid. After it had been rapidly cooled to 0–5 °C, a 5 mL solution containing 0.72 g (69 g/mol, 0.0105 mol) of sodium nitrite was added. Ehrlich reagent was used to detect the termination of diazotisation, and sulfamic acid was used to remove the residual nitrous acid. The coupling component 1-(2-ethylhexyl)-3-(2-ethylhexyl)-5-pyrazalone 3.08 g (308 g/mol, 0.01 mol) was dissolved in 100 mL ethanol and cooled to 0–10 °C. Diazonium salt solution was added to the above solution at a temperature below 10 °C over a 15–20 min period, with the pH level maintained at 8–9 using a sodium hydroxyl solution. The reaction mixture was stirred for a further 3 h. Pyrazolone-1 was precipitated by adjusting the pH of the coupling solution to 4.0 using acetic acid, then collected and dried to get 4.5 g crude product Pyrazolone-2. The crude Pyrazolone-2 was purified by flash chromatography on silica gel (eluting agent: petroleum ether and ethyl acetate). Characterization: yellow solid; yield: 87.8%; λ (max) = 416 nm (in decane); ε = 25335 L·mol^{-1}·cm^{-1}. IR (KBr, cm^{-1}): 3103.20 (Ar–H); 2958.40 (–CH$_3$); 2929.40 (–CH$_2$$^-$); 2863.30 (–CH$_3$); 1695.60 (C=O), 1617.00 (Ar); 1579.80 (Ar); 1521.90 (Ar); 1461.90 (–N=N–); 1379.20 (–CH$_3$); 1226.20 (C–N–C); 1174.50 (C–N–C); 1120.82; 1044.32; 891.32. ^{1}H NMR (500 MHz, CDCl$_3$): 13.51 (s, 1H, O–H), 7.92–7.90 (d, 2H, J = 9.0 Hz, Ar–H), 7.83-7.82 (d, 2H, J = 7.5, Ar–H), 7.45–7.36 (m, 5H, Ar–H), 3.60–3.52 (m, 2H, –NCH$_2$$^-$), 2.60–2.57 (t, 2H, J = 15.5, Ar–CH$_2$$^-$), 1.79 (m, 1H, –CH–), 1.67 (m, 2H, –CH$_2$$^-$), 1.23 (m, 10H, –CH$_2$$^-$), 0.92-0.88 (t, 3H, J = 15.0, –CH$_3$), 0.86-0.80 (m, 6H, –CH$_3$). ^{13}C NMR (500 MHz, CDCl$_3$): 158.65, 152.72, 149.92, 143.49, 130.93, 129.13, 124.73, 122.82, 115.80, 47.67, 38.45, 32.94, 30.62, 28.77, 26.21, 23.41, 14.12, 10.72. MS (MALDI-TOF) (DIF) M/Z (%): 516.30 (M–H)$^-$. Calc. for (C28H40N4O): 516.40.

4. Results and Discussion

The ^{1}H NMR spectra for the target dyes recorded in CDCl$_3$ are shown in Figure 2. The structures of dyes were confirmed by the presence of four distinct downfield signals characteristic of the aromatic rings. For Anthra-1–2, the structure of the anthraquinone core was confirmed by the presence of four distinct downfield signals ("H1", "H2", "H3", "H4"). The "H1", "H2", "H3", and "H4" protons of Anthra-1-2 exhibited one singlet, two multiplets and one doublet between 10.71 ppm and 6.93 ppm. For the substituents in the N-position, these two dyes displayed one common multiple peak "H5" between 3.22 and 3.14 ppm, corresponding to the methene protons adjacent to the imide group. The "H6" protons appeared as a singlet between 0.88 and 0.83 ppm. For the other pyrazolone dyes Pyrazolone-1–2, the "H1–4" and "H7–9" protons of Pyrazolone-1 exhibited three separate doublets, one triplet, and one multiplet between 8.00 ppm and 7.43 ppm. The "H1", "H2", "H3" and "H4" protons of Pyrazolone-2 exhibited three separate doublets and one multiplet between 7.92 ppm and 7.36 ppm. Two dyes: Pyrazolone-1–2, displayed one common multiple peak "H5" between 3.60 and 3.52 ppm corresponding to the methene protons adjacent to the imide group. Moreover, the common multiple peak "H6" of those two dyes represented the protons on the methyl group, and the "H6" protons existed between 0.78 and 0.92 equivalent to the twelve protons which were obtained. The chemical values confirmed the structural correctness of the four dyes.

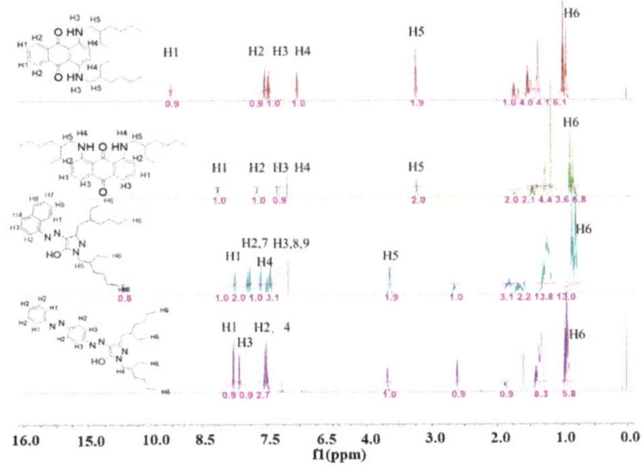

Figure 2. ^{1}H NMR spectra of four synthesized dyes.

4.1. Absorption Properties of Four Dyes

As can be seen in Figure 3, the uv-visible absorption spectra and data of all the dyes were recorded in decane and were found in the range of 416–650 nm, showing color variations from yellow to cyan. For the anthraquinone dyes Anthra-1–2, the maximum wavelength number (λ_{max}) value depended on the position of the alkyl-amino substituented on the anthraquinone ring. When two alkyl-amino groups were substituted on 1,4-sites, the λ_{max} was 650 nm, while the λ_{max} was 545 nm when they were substituted on 1,8-sites. For azo pyrazolone dyes Pyrazolone-1~2, the λ_{max} value depends on the conjugate rings. The λ_{max} of Pyrazolone-1, which is a mono-azo dye with a naphene group, is 416 nm. While Pyrazolone-2 is a bisazo dye with two benzene groups, the size of conjugate rings is a bit larger than Pyrazolone-1, thus the λ_{max} values (420 nm) are a bit longer than for Pyrazolone-1.

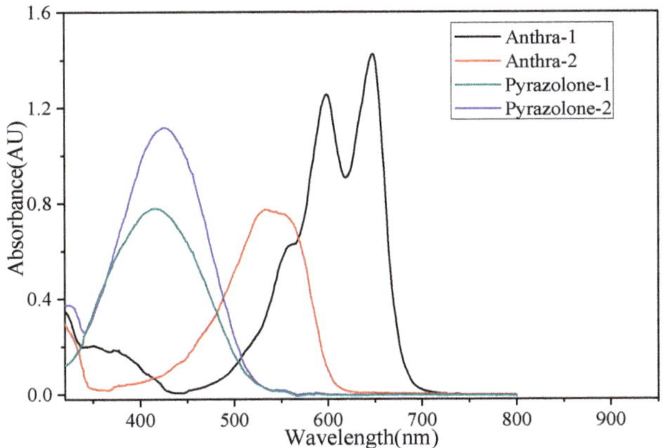

Figure 3. UV-visible absorption spectra of four dyes.

4.2. Photoelectric Response Properties of Dyes

The EFD cell was fabricated using normal processes. The switching behavior of the EFD cells was studied under a certain dc voltage. The transmission spectra of EFD displays were tested under different voltages. The results can be seen in Figure 4. For Pyrazolone-1 as an example, we can clearly see the transmission changes as the voltage rose to 20 V, showing that the oil film starts to break up at this voltage. By increasing the voltage from 0 V to 40 V, the transmission value of EFD cell at maximum wavelength (420 nm) increases from 20% to 70%, showing that the cell could switch from bright yellow color (under 0 V) to the most transparent or white color (under 40 V). The EFD display could not be changed to an obvious one transparent to the naked eye until the voltage reached 32 V. And the transmission curves did not change greatly under the higher voltage because of the saturation of the oil contraction. The micron photos of EFD displays at 0 V (the OFF state) and 32 V (the ON state) are shown in Figure 4. The aperture ratio was achieved by calculating the occupying area of oil and white area in Photoshop software, and the aperture ratio could reach as high as 68.5%–75.0%.

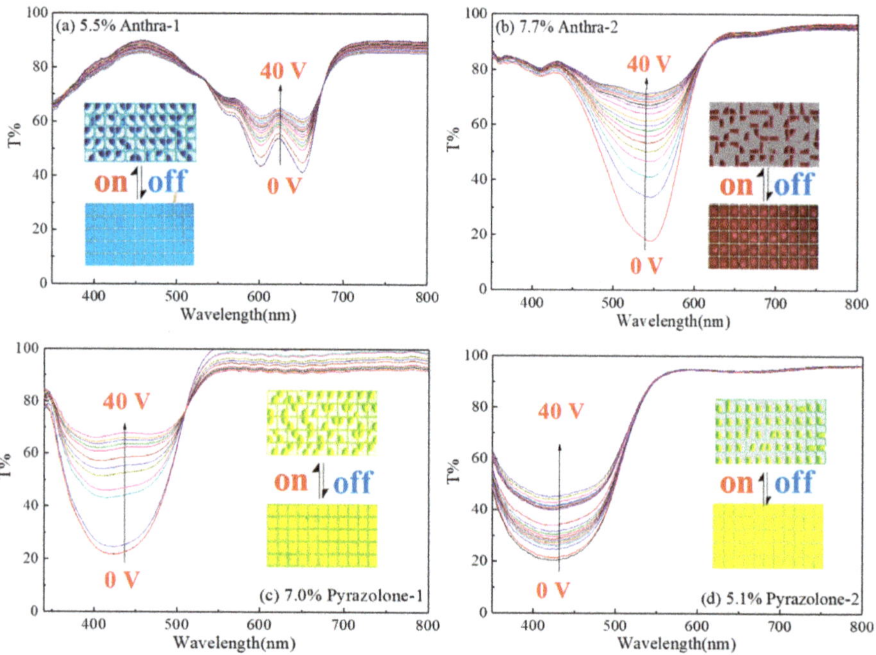

Figure 4. Transmission spectra of EFD display formulated by four formulated dyes under different voltages. (Insert: micron photos of pixels filled with different oils). (**a**) 5.5% Anthra-1, (**b**) 7.7% Anthra-2.

4.3. Photo-Stability Research of Dyes

The photo-stability test of EFD oils were evaluated following the standard method 60068-2-9. The aging test of the display was carried under accelerated sunlight conditions for 100 h at 45 °C, which can translate to several years of application lifetime under normal conditions [19].

The change of absorbance curves under different irradiation time is shown in Figure 5. The decrease of dye absorbance after a period of irradiation test depends on the degree of dye degradation. Generally, we can see that dye degradation of Anthra-2 and Pyrazolone-2 are higher than Anthra-1 and Pyrazolone-1. After 100h irradiation, only 1.2% of Anthra-1 dyes were decomposed. The good photo-stability of Anthra-1 was attributed to the substitution of –NH at 1,4-sites of anthrquinone structure, which could form two six-membered rings with two carbonyl groups. This special structure could strengthen the electron delocalization and decrease the electron density of potentially destabilizing groups, such as C–N and C–O bonds. In contrast, 23.9% of Anthra-2 decomposed after only 40 h of irradiation. The structure of Anthra-2 is a 1,8- sites of anthrquinone, while only one six-membered ring could be formed at the 1-site. As a result, the other substitution of –NH at the 8-site could not be stabilized and was prone to decomposing under irradiation. The initial anthraquinone dyes underwent a hydroxylation of the alkyl carbon atom adjacent to the left hand nitrogen atom. This resulted in a photo-bleached molecule with the original alkly group of the left hand alkl amine now substituted with the group –CHO [20].

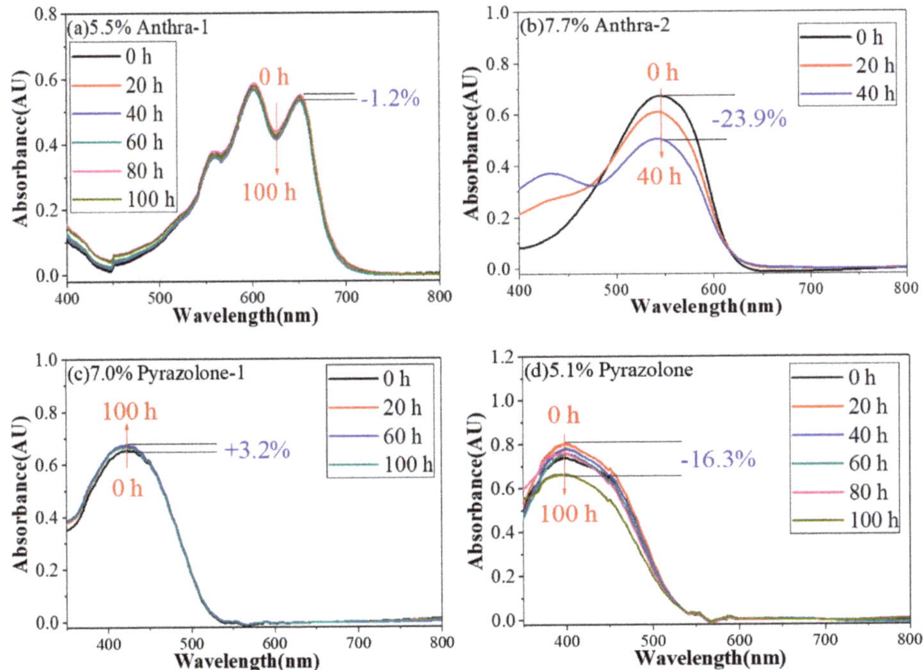

Figure 5. Photo-stability test results of four formulated dyes. (Irradiation condition: 0.55 W/m^2 (340 nm); Temperature 45 °C). (**a**) 5.5% Anthra-1, (**b**) 7.7% Anthra-2, (**c**) 7.0% Pyrazolone-1, (**d**) 5.1% Pyrazolone.

For azo pyrazolone dye Pyrazolone-1–2, it could be seen that after 20 h of irradiation, the absorption values of Pyrazolone-1 and Pyrazolone-2 increased by 3.2% and 10%, respectively. This abnormal phenomenon could be explained by the transformation of cis-tran isomerization of azo group caused by light. However, when the irradiation time was extended to 100 h, the absorption increase ratio R% of Pyrazolone-1 was maintained at around 3.2%, while the absorption of Pyrazolone-2 decreased to 83.7%, showing that 16.3% of Pyrazolone-2 was decomposed and few Pyrazolone-1 molecules were decomposed. This difference could also be explained by the difference of structures. Pyrazolone-2 is a bisazo dye, and the azo group adjacent to the pyrrazole structure could form a six-membered ring, while the other isolation azo group could be easily destroyed by strong photo irradiation. Pyrazolone-1 is a monoazo dye, meaning the azo group adjacent to pyrrazole structure could form a six-membered ring and was stabilized. Thus, it is much more stable than Pyrazolone-2.

The color coordinates (x, y) of colored EFD cell in CIE1931 color space and color deviation ΔE were monitored along with the accelerating irradiation test. The results could be seen in Figure 6 and Table 1. With the exposure of irradiation time, the color coordinates (x, y) of the colored EFD cell in the CIE1931 color space changed, showing changes of the color. Among the four colored EFD oils, the color coordinates (x, y) of Anthra-2 had the greatest change from (0.38, 0.20) to (0.45, 0.32), showing that the color changed from bright magenta to brown. The ΔE of Anthra-2 was found to be as high as 130.5 after 100 h of irradiation. The color coordinates (x, y) of Pyrazolone-1 changed from (0.462, 0.468) to (0.460, 0.468), showing that the color did not change anymore. The ΔE of Pyrazolone-1 was 2.0, showing a small color change which can hardly be distinguished by the naked eye. The color coordinates (x, y) of Anthra-1 and Pyrazolone-2 changed from (0.172, 0.215), (0.478, 0.464) to (0.183, 0.228), (0.479, 0.466), respectively. The ΔE of Anthra-1 and Pyrazolone-2 were 7.7 and 3.7, respectively, showing a slight color change. These performances are highly relevant to their structure stability. Structure of Anthra-2 is the least stable and its color changed a great deal, while structure of Pyrazolone-1 is the most stable and its color changed minimally.

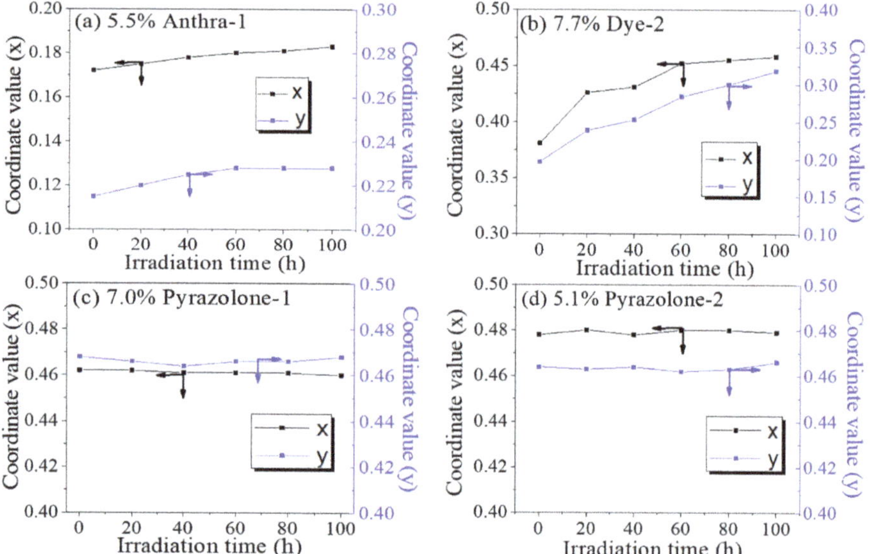

Figure 6. Variation of color coordinates (x, y) of oils in CIE1931 color space and ΔE. (Irradiation condition: 0.55 W/m^2 (340 nm); Temperature 45 °C). (**a**) 5.5% Anthra-1, (**b**) 7.7% Dye-2, (**c**) 7.0% Pyrazolone-1, (**d**) 5.1% Pyrazolone-2.

Table 1. Color change (ΔE) of four formulated dyes after accelerated irradiation.

Dye	Before Irradiation			After Irradiation			ΔE
	L	*a*	*b*	*L*	*a*	*b*	
Anthra-1	174.9	−45.5	−123.6	171.2	−43.5	−117.1	7.7
Anthra-2	140.8	181.0	−68.2	160.7	113.2	41.6	130.5
Pyrazolone-1	236.2	14.0	210.5	236.3	12.3	209.3	2.0
Pyrazolone-2	233.3	29.7	221.4	231.9	28.1	224.6	3.7

(Irradiation condition: 0.55 W/m^2 (340 nm); Irradiation time: 100 h; Temperature 45 °C).

4.4. Effect of Irradiation on Backflow Property of EFD Devices

The tendency of the oil to flow back into the pixel at a constant voltage is called "backflow". The reduction oil-level contraction affects pixel aperture for the backflow of oil, which has an impact on the ability of the switch and the power consumption of the display. Since a static reading is the mode of the operation for an EFD display, a low "backflow" property oil will lead to the decrease of the refresh rate for the display. Thus, the power consumption can be reduced because of its scaling with the refresh rate.

In Figure 7, it can be seen that before irradiation, the backflow time of four oils were around 50–400 s under 30 V DC (Direct Current) voltage, which is rather long and acceptable. The faster refresh speed for such a low backflow property will lead to a static image in an EFD display. A long backflow time can be attributed to the non-polarity of dye structures. The discrimination of backflow time for four oils is due to the difference of the non-polarity of their structures. When EFD cells underwent irradiation for a certain period, the backflow time of four dyes deteriorated sharply. For Anthra-2, the cell cannot be open after only 20 h of irradiation. This result might cause a dye's decomposition. When Anthra-2 was decomposed under irradiation, small fragments were released into oil, resulting in the rise of oil conductivity, thus the backflow time was extremely short and even pixels could not be opened. For Anthra-1 and Pyrazolone-1, in spite of the least dye decomposition, the backflow time also decreased from ~400 s and ~300 s to ~15 s and ~100 s, respectively after only 20 h of irradiation, and cells could not opened after 60 h of irradiation. These results showed that the effect of irradiation on the backflow property of EFD devices is not only relevant to a dye's structure, but also relates to the other materials, such as pixel wall materials and fluoropolymer materials. Any decomposition of these materials will release small high polarity fragments into oil, causing a rise in oil conductivity, and a decrease of the backflow time of EFD cell. This effect will be studied in-depth by our group, so it will not be extensively discussed in this paper.

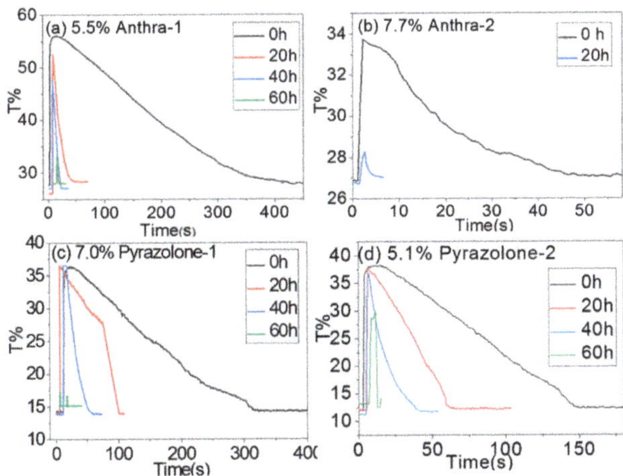

Figure 7. Backflow properties of EFD devices with dyes under different irradiation times. (Irradiation condition: 0.55 W/m^2 (340 nm); Temperature 45 °C). (**a**) 5.5% Anthra-1, (**b**) 7.7% Anthra-2, (**c**) 7.0% Pyrazolone-1, (**d**) 5.1% Pyrazolone-2.

5. Conclusion

In summary, we have synthesized four EFD oil materials based on anthraquinone and azo pyrazolone dyes. The absorption, electrical-optical response, and photo-stability of these dyes were researched in detail. The dyes have good electrical-optical response properties due to their low polarity of whole molecules. However, the photo-stability was highly dependent on the structures. High polarity and high electronic density of the chemical bond are the photo-sensitive groups which are potentially destroyed under photo irradiation. Six-membered rings formed by hydrogen bonds could play an important role as a "protective structure", and they can highly increase the photo-stability property of these dyes. Irradiation could sharply decrease the backflow time of existing EFD devices, which is not only caused by the decomposition of oil materials, but also may be caused by the other materials in the pixels.

Author Contributions: Conceptualization, Y.D.; methodology, Y.D.; software, Y.D.; validation, Y.D. and S.L.; formal analysis, S.L.; investigation, H.J.; resources, Y.D.; data curation, H.J.; writing—original draft preparation, Y.D. and S.L.; writing—review and editing, Y.D.; visualization, D.Y.; supervision, B.T.; funding acquisition, G.Z. 'All authors have read and agreed to the published version of the manuscript.

Funding: This work was supported by National Key R&D Program of China (No. 2016YFB0401501), National Natural Science Foundation of China (Grant No. U1601651), Program for Changjiang Scholars and Innovative Research Team in University (IRT13064), Program for Science and technology project of Guangdong Province (Nos. 2018A050501013), National Natural Science Foundation of Guangdong, China (2018A0303130059), Science and Technology Program of Guangzhou (No. 201904020007, 2019050001), Guangdong Provincial Key Laboratory of Optical Information Materials and Technology (Grant No. 2017B030301007), MOE International Laboratory for Optical Information Technologies and the 111 Project.

Conflicts of Interest: The authors declare no conflict of interest.

References

1. Hayes, R.A.; Feenstra, B.J. Video-speed electronic paper based on electrowetting. *Nature* **2003**, *425*, 383–385. [CrossRef] [PubMed]

2. Tang, B.; Groenewold, J.; Zhou, M.; Hayes, R.A.; Zhou, G.G. Interfacial electrofluidics in confined systems. *Sci. Rep.* **2016**, *6*, 26593. [CrossRef] [PubMed]

3. Shui, L.; Hayes, R.A.; Jin, M.; Zhang, X.; Bai, P.; van den Berg, A.; Zhou, G. Microfluidics for electronic paper-like displays. *Lab Chip* **2014**, *14*, 2374–2384. [CrossRef] [PubMed]

4. Riahi, M.; Brakke, K.A.; Alizadeh, E.; Shahroosvand, H. Fabrication and characterization of an electrowetting display based on the wetting–dewetting in a cubic structure. *Opt. Int. J. Light Electron Opt.* **2016**, *127*, 2703–2707. [CrossRef]

5. Wang, M.; Guo, Y.; Hayes, R.; Liu, D.; Broer, D.; Zhou, G. Forming spacers in situ by photolithography to mechanically stabilize electrofluidic-based switchable optical elements. *Materials* **2016**, *9*, 250. [CrossRef] [PubMed]

6. Smith, N.R.; Hou, L.; Zhang, J.; Heikenfeld, J. Fabrication and demonstration of electrowetting liquid lens arrays. *J. Disp. Technol.* **2009**, *5*, 411–413. [CrossRef]

7. Heikenfeld, J.; Smith, N.; Dhindsa, M.; Zhou, K.; Kilaru, M.; Hou, L.; Zhang, J.; Kreit, E.; Raj, B. Recent progress in arrayed electrowetting optics. *Opt. Photonics News* **2009**, *4*, 20–26. [CrossRef]

8. Heikenfeld, J.; Zhou, K.; Kreit, E.; Raj, B.; Yang, S.; Sun, B.; Milarcik, A.; Clapp, L.; Schwartz, R. Electrofluidic displays using Young-Laplace transposition of brilliant pigment dispersions. *Nat. Photonics* **2009**, *3*, 292–296. [CrossRef]

9. Van De Weijer, M.M.H.; Massard, R.; Hayes, R.A. Electrowetting Elements. U.S. Patent 8,980,141, 17 March 2011.

10. Ishida, M.; Shiga, Y.; Takeda, U.; Kadowaki, M. Ink Containing Anthraquinone Based Dye, Dye Used in the Ink, and Display. U.S. Patent 8,999,050, 7 April 2013.

11. Shiga, Y.; Takeda, U.; Ichinosawa, S.; Ishida, M. Ink Containing Heterocyclic azo Dye, and Dye for Use in Said Ink. U.S. Patent 8,747,537, 10 June 2014.

12. Shiga, Y.; Ishida, M. Pyrazole Disazo Dye and Ink Containing the Dye. U.S. Patent 8,143,382, 27 March 2012.

13. Chiang, Y.; Chao, Y. Synthesis of dis-azo black dyes for electrowetting displays. *Mater. Sci. Eng. B* **2012**, *177*, 1672–1677. [CrossRef]

14. Farrand, L.D.; Smith, N.; Corbett, A. and Lawrence, A.; Merck Patent GmbH. Electrowetting Fluids. U.S. Patent 20150355456 A1, 10 December 2015.

15. Chiang, Y.F.; Chao, Y.C. Synthesis and Application of Oil-Soluble Red Dyes Derived from p-n-Alkyl Aniline. *Mater. Sci. Appl.* **2014**, *5*, 485–490. [CrossRef]

16. Kato, T.; Higuchi, S.; Fukushige, Y.; Jimbo, Y.; Sasaki, D. Colored Composition for Electrowetting Display, Image Display Structure, and Electrowetting Display Device. U.S. Patent 9,494,789, 15 November 2014.

17. Lee, P.T.; Chiu, C.W.; Chang, L.Y.; Chou, P.Y.; Lee, T.M.; Chang, T.Y.; Wu, M.T.; Cheng, W.Y.; Kuo, S.W.; Lin, J.J. Tailoring Pigment Dispersants with Polyisobutylene Twin-Tail Structures for Electrowetting Display Application. *ACS Appl. Mater. Interfaces* **2014**, *6*, 14345–14352. [CrossRef]

18. Lee, P.T.; Chiu, C.W.; Lee, T.M.; Chang, T.Y.; Wu, M.T.; Cheng, W.Y.; Kuo, S.W.; Lin, J.J. First Fabrication of Electrowetting Display by Using Pigment-in-Oil Driving Pixels. *ACS Appl. Mater. Interfaces* **2013**, *5*, 5914–5920. [CrossRef] [PubMed]
19. Deng, Y.; Jiang, H.; Ye, D.; Zhou, R.; Li, H.; Tang, B.; Jin, M.; Li, N.; Guo, Y.; Zhou, G. Synthesis and application of an alkylated pyrazole-based azo dye for Electro-fluidic display. *J. Soc. Inf. Disp.* **2018**, *26*, 369–375. [CrossRef]
20. Van de Weijer, W.; Melanie, M.; Massard, R.; Hayes Robert, A. Improvements in Relation to Electrowetting Elements. U.S. Patent WO2010031860 A2, 25 March 2010.

 *micromachines* 

Article

Aperture Ratio Improvement by Optimizing the Voltage Slope and Reverse Pulse in the Driving Waveform for Electrowetting Displays

Zichuan Yi [1], Wenyong Feng [2,3,*], Li Wang [3,*], Liming Liu [1], Yue Lin [1,2], Wenyao He [2], Lingling Shui [1,2], Chongfu Zhang [1], Zhi Zhang [1] and Guofu Zhou [2,3]

[1] College of Electron and Information, University of Electronic Science and Technology of China, Zhongshan Institute, Zhongshan 528402, China; yizichuan@163.com (Z.Y.); liulmxps@126.com (L.L.); linyuesky@gmail.com (Y.L.); Shuill@m.scnu.edu.cn (L.S.); cfzhang@uestc.edu.cn (C.Z.); zz001@zsc.edu.cn (Z.Z.)

[2] Institute of Electronic Paper Displays, South China Academy of Advanced Optoelectronics, South China Normal University, Guangzhou 510006, China; hwy956005579@outlook.com (W.H.); guofu.zhou@m.scnu.edu.cn (G.Z.)

[3] Shenzhen Guohua Optoelectronics Tech. Co., Ltd., Shenzhen 518110, China

* Correspondence: wenyong.feng@guohua-oet.com (W.F.); creekxi@163.com (L.W.); Tel.: +86-0755-2941-5855 (L.W.)

Received: 3 November 2019; Accepted: 5 December 2019; Published: 7 December 2019

Abstract: Electrowetting display (EWD) performance is severely affected by ink distribution and charge trapping in pixel cells. Therefore, a multi structural driving waveform is proposed for improving the aperture ratio of EWDs. In this paper, the hysteresis characteristic (capacitance–voltage, C-V) curve of the EWD pixel is tested and analyzed for obtaining the driving voltage value at the inflection point of the driving waveform. In the composition of driving waveform, a voltage slope is designed for preventing ink dispersion and a reverse pulse is designed for releasing the trapped charge which is caused by hysteresis characteristic. Finally, the frequency and the duty cycle of the driving waveform are optimized for the max aperture ratio by a series of testing. The experimental results show that the proposed driving waveform can improve the ink dispersion behavior, and the aperture ratio of the EWD is about 8% higher than the conventional driving waveform. At the same time, the response speed of the driving waveform can satisfy the dynamic display in EWDs, which provides a new idea for the design of the EWD driving scheme.

Keywords: electrowetting display; aperture ratio; driving waveform; hysteresis characteristic; ink distribution; response speed

---

1. Introduction

In recent years, low-power and reflective displays which are readable in the sun are favored by many scientific researchers [1]. As a kind of reflective display technology, the electrowetting display (EWD) is receiving more and more attention [2], and its initial products have been successfully applied to various fields [3]. However, a stable multi-grayscale video display of EWDs has still not been implemented.

In 1981, researchers proposed the EWD technology [4]. It can display grayscale by controlling the movement of the colored ink. In the past few decades, various driving systems and image processing methods based on multi-level grayscale dynamic EWDs have greatly improved display quality [5,6]. We have also proposed a driving system for multi-grayscale display in an EWD based on a field programmable gate array (FPGA) chip, which provides an important reference for optimizing the dynamic display of EWDs [7]. In the field of EWD driving waveform, the driving method based on amplitude modulation (AM) or pulse width modulation (PWM) has been proposed [8], and the driving voltage is used to control the shrinkage of the ink droplets. However, the ink in the EWD pixel shows a phenomenon of hysteresis, which has a negative effect on optical performance. The PWM compensates for the hysteresis to some extent, but the PWM means a reduction in response speed and it can reduce the effective frame for the video display, and the power consumption of the EWD grayscale display is also increased by using the PWM at the same time. In addition, a dynamic contact angle model is established based on the molecular dynamics theory; the shape evolution of droplets has been studied under different direct current (DC) driving conditions. In addition, the influence of liquid interface resonance on contact angle has also been studied under alternating current (AC) driving conditions [9]. However, there are only two types of driving waveforms for driving droplets: stepped DC driving signal and sinusoidal AC driving signal. The problem of ink splitting has been also proposed during coupling AC-common driving process [10]. A driving modulation scheme was used to improve the ink dispersion performance to a certain extent, but the ink showed instability during the driving process. In other driving modes, sinusoidal, bipolar, and single-pulse are used to drive electrowetting liquid lenses respectively, and it was found that the positive and negative polarities of the driving voltage have a significant effect on the charge trapping of liquid lenses [11]. This provides a possible direction for the optimization of the driving waveform for EWDs. Based on the EWD micro-space pixel units, it has been found that the ink motion shape can be controlled by applying driving voltage with different rising speed [12]. The EWD reflectivity was improved using this method, but the response time of the driving waveform became longer.

In order to improve the aperture ratio of EWD pixels and the shrinkage form of the ink, the relationship among driving voltage, ink shape, pixel aperture ratio, and response time are studied in this paper. Then, a driving waveform is proposed for EWDs to improve the display effect according to the motion behavior of the ink and the structure characteristics of the EWD. Compared with the conventional driving waveform, the proposed driving waveform can improve the performance of charge trapping, ink dispersion, aperture ratio, and response time.

2. Principle of EWD Driving Waveform System

2.1. Electrowetting Equivalent Circuit Model

Grayscale display is realized in EWDs by applying voltage to control the movement of colored ink droplets. Its essence is an optical switch, which has excellent grayscale display characteristics. A typical EWD pixel structure is shown in Figure 1a. The orange part is the pixel wall, which can control the range of ink movement. From the top down, the ink is flat, and the state is also the lowest energy state.

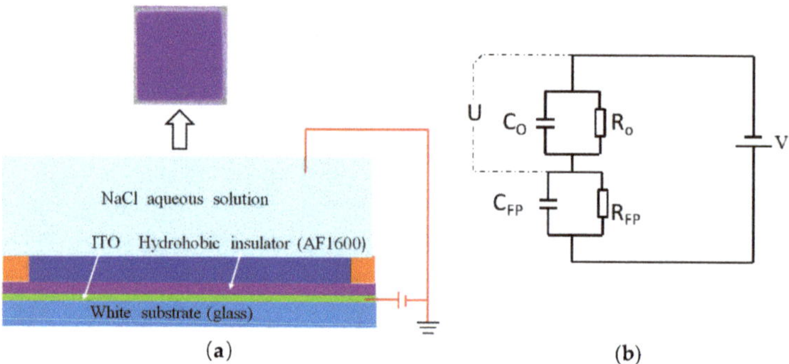

Figure 1. The electrowetting display (EWD) pixel structure and its equivalent circuit. (**a**) The EWD pixel structure without applied voltage; (**b**) simplified equivalent circuit diagram of an EWD pixel unit.

The voltage is applied between the upper and lower plates which are made up of indium tin oxide (ITO). According to the Lippmann–Young equation, with the increase of the applied electric field, the surface tension among the insulating layer, ink film, and water can be increased, and the contact angle between the interfaces will also increase. The original balance is broken by the electric field force which is generated by the voltage difference; the ink is squeezed away by the water, and the water contacts the surface of the hydrophobic insulation layer, then, the white base plate is exposed. The aperture ratio (white area ratio) W_A of a pixel is defined as Equation (1).

$$W_A = 1 - \left(\frac{S_{ink}(V)}{S_{pix}} \right) \times 100\%　\tag{1}$$

In Equation (1), S_{ink} and S_{pix} are defined as the surface area of the ink in a single pixel and the surface area of the entire pixel, respectively, and V represents the voltage applied on the pixel in the EWD. In addition, the area of the pixel wall is ignored in the calculation of the aperture ratio. The displacement of the ink shrinkage is determined by the applied voltage [13,14]. Theoretically, the radius of ink shrinkage is directly related to the contact angle, and the contact angle θ follows the Lippmann–Young equation.

$$\cos\theta = 1 - \frac{CV^2}{2\gamma_{OW}}　\tag{2}$$

In Equation (2), C is the capacitance of a single pixel unit area. V is the voltage applied on the pixel unit, and γ_{OW} is the ink–water interfacial tension. The capacitance–voltage (C-V) curve provides an important parameter for driving EWDs and is an important basis for the design of driving waveform voltage.

For the equivalent circuit model of the micro-space pixel unit, the ink droplets are driven under an alternating electric field, and the ink and pixel wall in the same dielectric layer are treated as a combined loop. The photoresist material of the pixel wall has low conductivity, and the simplified equivalent circuit diagram is shown in Figure 1b [15]. C_0 is the ink capacitance, R_0 is the ink resistance, C_{FP} is the dielectric layer capacitance, R_{FP} is the dielectric layer resistance, and U is the effective voltage applied to the ink in the pixel. Based on the equivalent circuit model, the resistance, capacitance of the insulating layer, and ink follow Ohm's law; then we can get their expressions of a single pixel.

Dielectric layer resistance is shown in Equation (3).

$$R_{FP} = \frac{d}{S\delta_{FP}}　\tag{3}$$

In Equation (3), d represents the thickness of the dielectric layer; S represents the area of a single pixel unit; δ_{FP} represents the dielectric constant of the hydrophobic layer. The ink resistance is shown in Equation (4).

$$R_0 = \frac{H}{S\delta_I}　\tag{4}$$

In Equation (4), H represents the thickness of the ink; δ_I represents the dielectric constant of the ink. The dielectric layer capacitance is shown in Equation (5).

$$C_{FP} = \varepsilon_0\delta_{FP}S/d　\tag{5}$$

In Equation (5), ε_0 is the vacuum dielectric constant. The ink capacitance C_I is shown in Equation (6).

$$C_I = \varepsilon_0\delta_I S/H　\tag{6}$$

In Equation (6), δ_I represents the dielectric constant of the ink; ε_0 is the vacuum dielectric constant. According to Equations (3) and (4), the effective resistance R is shown in Equation (7).

$$R = \frac{1}{S}\left(\frac{d}{\delta_{FP}} + \frac{H}{\delta_I}\right) \tag{7}$$

According to Equations (5) and (6), the single pixel cell effective capacitance C(H) is shown in Equation (8).

$$C(H) = \varepsilon_0 S\left(\frac{\delta_{FP}}{H} + \frac{\delta_I}{d}\right) \tag{8}$$

The critical starting voltage V_{swith} of a single EWD pixel unit can be estimated by Equation (9).

$$V_{swith} = \sqrt{\frac{2\pi^2 \gamma_{ow}}{C(H)S}} \tag{9}$$

The insulating film Teflon used in the EWD pixel cell structure has a correspondence relationship with the threshold voltage [16].

$$\cos\theta_\alpha - \cos\theta_0 = \frac{\delta\varepsilon_0 U_{EW}^2}{2d\gamma_{ow}} \tag{10}$$

In Equation (10), θ_α is the Lippmann contact angle; θ_0 is the static contact angle; δ is the dielectric constant of a single pixel. The relationship between the applied voltage U_{EW} and the film thickness d of the insulating layer can be derived in Equation (11).

$$U_{EW} = \sqrt{\frac{2d\gamma_{ow}}{\delta\varepsilon_0}(\cos\theta_\alpha - \cos\theta_0)} \sim \sqrt{d} \tag{11}$$

The voltage of the insulating layer is proportional to the film thickness. The insulating layer is easily electrically broken when the film thickness of the insulating layer is too thin, so the parameters of the film thickness of the insulating layer should be considered in the design of the driving waveform.

2.2. Analysis of Ink Distribution in a Pixel

In the process of driving an EWD, it is ideal to drive the ink to one corner of a pixel, as shown in Figure 2a, so we can get a max value of the aperture ratio at this time. However, once a conventional square driving waveform is applied, the ink may shrink to two or more corners, as shown in Figure 2b. The shape of the ink dispersion affects the EWD aperture ratio directly. As shown in Figure 3, one ink droplet A is divided into two droplets: B and C. Obviously, the covered area of the A droplet on the substrate S_A is smaller than the sum area of S_B and S_C and the aperture ratio becomes smaller when the ink is divided into two parts.

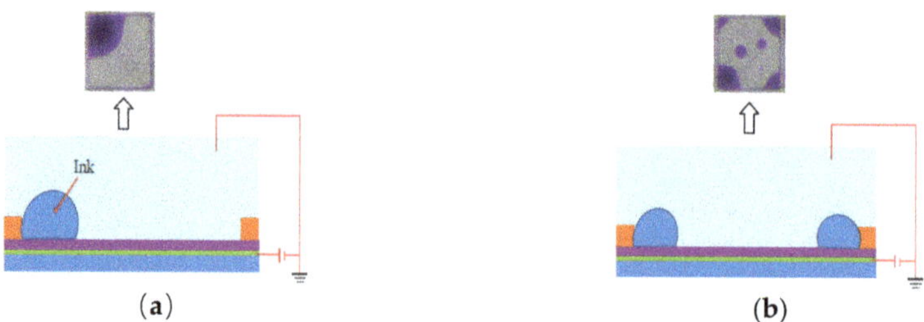

Figure 2. Ink film distribution state when the pixel is driven by the driving waveform. (**a**) Ink shrinks to one corner; (**b**) ink shrinks to four corners.

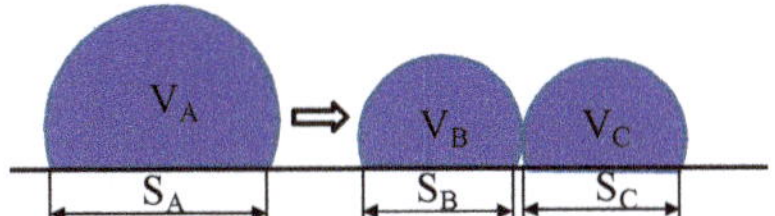

Figure 3. Ink splitting diagram.

In an EWD pixel, the ink shape mainly has four shapes, as shown in Figure 4. Figure 4a is an ideal ink shape, the ink is all shrunk to one corner in a pixel. The maximum aperture ratio can be obtained, and the aperture ratio can reach 73.9%. However, if a conventional driving waveform is applied to the EWD panel, the ink may be dispersed into two parts as shown in Figure 4b, or dispersed into three parts as shown in Figure 4c, or dispersed into four parts as shown in Figure 4d. Obviously, the pixel can reach a max aperture ratio value when the ink distribution is stable. The aperture ratio values of different ink film distributions are shown as follows: 73.9%, 62.1%, 61.6%, and 60.7%, respectively. So, the greater the number of ink dispersions, the smaller the aperture ratio in the EWD pixel.

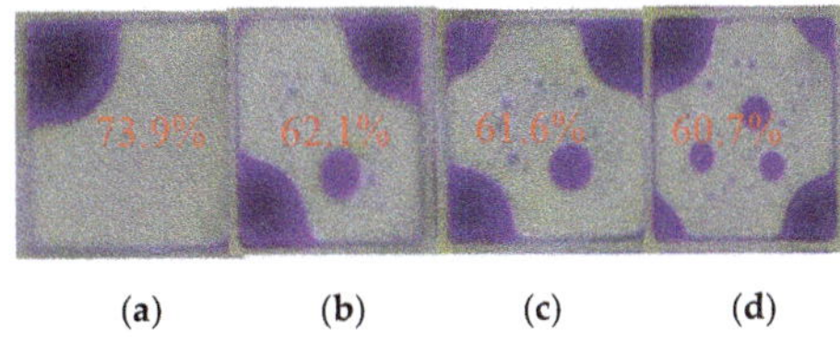

(**a**)　　　　(**b**)　　　　(**c**)　　　　(**d**)

Figure 4. The ink distribution state and the corresponding aperture ratio with the action of the driving waveform. (**a**) The ink is all shrunk to one corner and its aperture ratio value is 73.9%. (**b**) The ink is dispersed into two parts and its aperture ratio value is 62.1%. (**c**) The ink is dispersed into three parts and its aperture ratio value is 61.6%. (**d**) The ink is dispersed into four parts and its aperture ratio value is 60.7%.

3. Driving Waveform Design

3.1. Testing System

In order to test the effects of the driving waveform, the experimental setup is designed as shown in Figure 5. In the tested EWD panel (2.7 inches diagonally), the size of a single pixel grid is 150 μm × 150 μm, and an entire panel contains 78,408 independent pixels. The width and height of the pixel wall are 15 μm and 5.6 μm, respectively, and the thickness of the insulating layer is 1 μm. The height between the ITO substrate and the upper cover in the pixel grid is 75 μm. The solvent C10H22 is used as a colored ink whose molecular concentration is 10 wt%, and the electrolyte is a sodium chloride solution whose concentration is 1.4 mol/L. The thickness of the spin-coated ITO glass substrate and the surface panel are 1.1 mm and 1.7 mm, respectively, and the impedance is 100 Ω/Sq. Teflon AF1600 is spin-coated on the surface of the glass substrate as a hydrophobic insulating layer, and the colored ink is injected into the pixel grid at a low speed (1 mm/s) by the grating filling method. Finally, the pixel unit is edge-sealed using a pressure sensitive element.

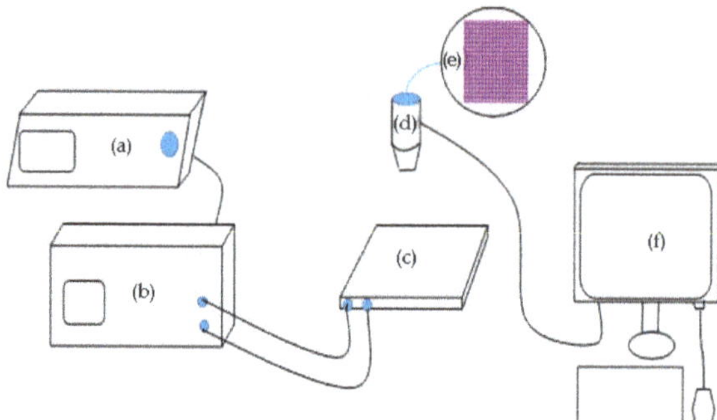

Figure 5. Optical testing system for EWDs. (**a**) AFG3052 arbitrary function generator; (**b**) ATA2022H high voltage amplifier; (**c**) testing board; (**d**) microscope; (**e**) pixels in an EWD; (**f**) computer.

An Agitek AFG-3052C function generator and an Agitek ATA-2022H high voltage amplifier were used as driving devices. The driving waveform was edited by the PC and transmitted to an arbitrary function generator. The high voltage amplifier can amplify the driving waveform and output the driving voltage to the EWD. The process of ink breakage is recorded by a camera and the video format is saved, then the EWD aperture ratio data is calculated by an image processing software. The system is shown in Figure 5.

3.2. The Driving Waveform Structure

In the design of the driving waveform, the C-V curve provides important parameters for the driving waveform design. The C-V curve of a typical EWD panel is shown in Figure 6, which can be tested by the optical testing system in Figure 5. The driving voltage steps from 0 V to 30 V, and its speed is 4 V/s. At each step voltage, a high-speed camera is used to record the real-time image of the EWD pixel and calculate the aperture ratio. In Figure 6, the threshold voltage of the ink rupture whose value is 15 V can be observed clearly. The capacitance of the pixel increases sharply when the driving voltage is higher than the threshold voltage, and the amount of the charge at the EWD three-phase contact line increases rapidly. However, the optical response is not linear with the change of the driving voltage, which is a significant hysteresis between the rising and falling of the C-V curve.

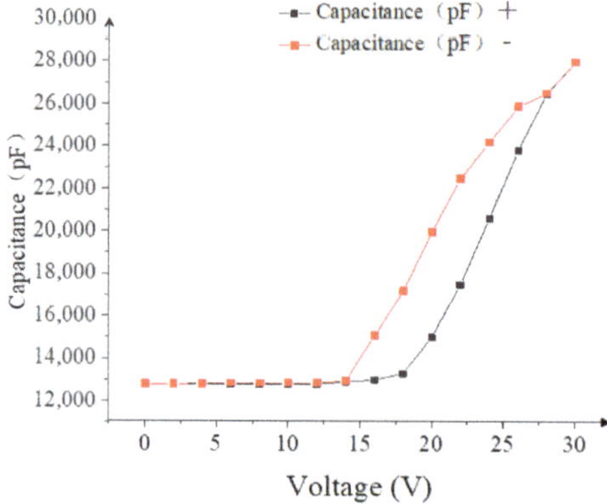

Figure 6. The relationship between the capacitance and the driving voltage in an EWD.

In addition, the threshold voltage for ink film reformation (red line) is lower than the threshold voltage for ink film dispersion (black line). According to the C-V curve, a large charge is rapidly accumulated in the hydrophobic insulator, which results in a sudden change of the electric field force which is likely to lead to ink dispersion when the capacitance value of the EWD pixel increases rapidly.

In order to avoid ink dispersion, we designed a new driving waveform with three stages. In the first stage the driving voltage starts from 0 V, and a reverse electrode pulse voltage of several milliseconds is applied to remove the electric charge which is trapped in the hydrophobic insulating layer. Hence, the polarization and the hysteresis phenomenon are avoided. The step voltage in this stage is raised to the threshold voltage (15 V) which is shown in the C-V curve. At the second stage a driving voltage with a slope of 0.4 V/s is applied, and the ink film is prevented from being dispersed during the rupture of the ink film, so the aperture ratio of the EWD can be kept at a stable value. During the third stage the duty cycle of the driving waveform is adjusted to keep the driving voltage at a high level to maintain the shrinkage state of the ink. A driving waveform with a rising slope for improving the ink dispersion phenomenon has been proposed [12], as shown in Figure 7a. The voltage rising time of the driving waveform is t2–t1 slower than the proposed driving waveform in this paper, as shown in Figure 7b.

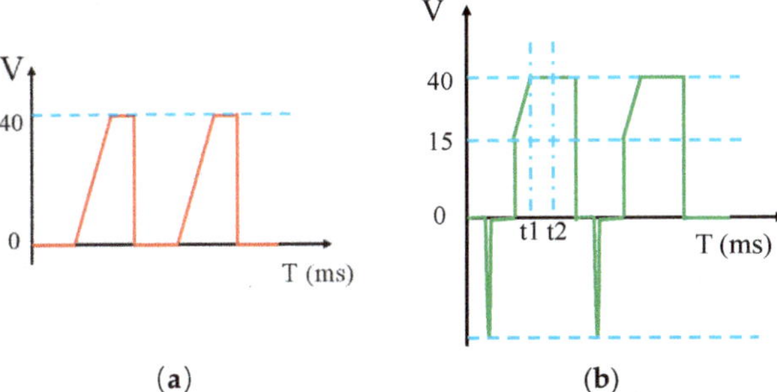

Figure 7. The structure of driving waveforms. (**a**) The driving waveform with a rising slope. (**b**) The proposed driving waveform in this paper.

The wetting mechanism of EWDs is related to the properties of ions in liquid solution. Because of defects in dielectric layer, ions in liquid can be bound by the dielectric layer easily, and the binding strength of the negative ion is greater than that of positive ion. Therefore, a large number of negative ions are bound by the dielectric layer when the positive voltage is applied. However, the negative ion cannot be released immediately when the applied voltage drops to 0 V, which results in hysteretic response of EWDs. According to the behavior of the ion which is trapped in dielectric layer, charge trapping in wetting dielectric layer can be controlled by the frequency of the driving waveform, reverse electrode driving mode and duty cycle coefficient (K) of the driving waveform.

4. Experimental Results and Discussion

4.1. The Frequency of the Driving Waveform

The frequency of the driving waveform depends on two factors: the viscosity of the liquid and the charged ions in the hydrophobic insulating layer. The charged ion leads to the hysteresis effect of ink shrinkage and spread. In addition, the higher the ink viscosity, the higher the driving voltage or lower driving frequency. The conversion time of the charged ion can be shortened when the driving waveform has a high frequency, but there is not enough time to remove the charged ion trapped in the insulating layer, which leads to the accumulation of ions in the hydrophobic insulating layer; this is the main factor which leads to the hysteresis phenomenon of EWDs. Thus, the setting of the best frequency directly affects the performance of the driving waveform.

In the testing process, the frequency of the driving waveform must higher than 25 Hz, because the flicker can be discerned by human eyes when the frequency is lower than 25 Hz [7]. Therefore, the conventional driving waveform and the proposed driving waveform in this paper are tested with different driving frequencies (30–90 Hz), which are higher than 25 Hz. The experimental results are shown in Figure 8. The aperture ratio is increased at the beginning, and then, it gradually decreases in the process of driving waveform frequency change from 30 Hz to 90 Hz; the highest aperture ratio is about 60 Hz. Hence, the change of ink shrinkage is related to the frequency of the driving waveform, which determines the release rate of charged ions. The aperture ratio value is smaller when the frequency is lower.

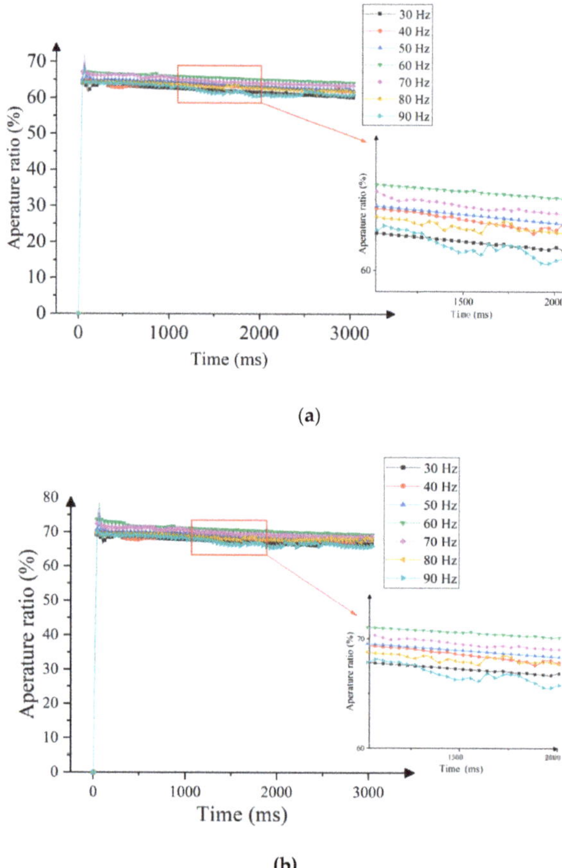

(a)

(b)

Figure 8. The relationship between the frequency and pixel aperture ratio. (**a**) Conventional driving waveform. (**b**) The proposed driving waveform in this paper.

In addition, the ink motion is mainly affected by charged ions which bind in dielectric layer, ink viscosity, and friction. However, the ink aperture ratio fluctuates when the ink motion lags behind the conversion of the driving waveform, which can lead to the decrease of the average aperture ratio, as shown in Figure 9. The aperture ratio of the pixel and the number of charges bound in the dielectric layer are decreased when the frequency is increased gradually. At the same time, the charged ion accumulated at the three-phase contact line is decreased, but the change speed of the charged ion cannot keep up with the voltage polarity conversion. In addition, the oscillation amplitude of the ink can be decreased when the frequency of the driving waveform is increased. So, T ≈ 1/60 s is the key cycle for the stable shrinkage of the ink in the EWD, as shown in Figure 9.

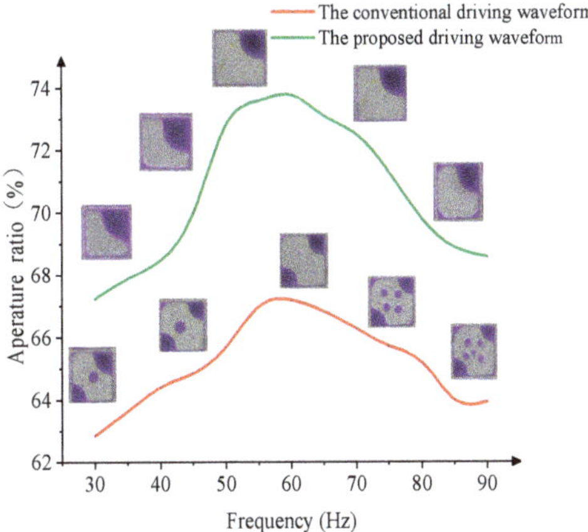

Figure 9. The relationship between aperture ratio (ink distribution) and the frequency of the driving waveform.

4.2. The Duty Cycle of the Driving Waveform

The aperture ratio of EWDs is determined by the voltage conversion frequency of the driving waveform. Nevertheless, the duty cycle of the driving waveform is another important factor for the performance of EWDs. In this paper, the experimental results show that the proposed driving waveform with a modulated duty cycle has better performance. Compared with the conventional driving waveform, the charged ion's behavior can be better controlled and the pixel aperture ratio can be improved by optimizing the duty cycle. In Figure 10, with the same driving voltage and key period ($T \approx 1/60$ s), the performance of the conventional driving waveform and the proposed driving waveform in this paper are compared in respect to the different duty cycle coefficients. Obviously, the two kinds of driving waveform have a maximum aperture ratio when the duty cycle coefficient is $K \approx 0.65$. So, the duty cycle of the proposed driving waveform is set as 0.65.

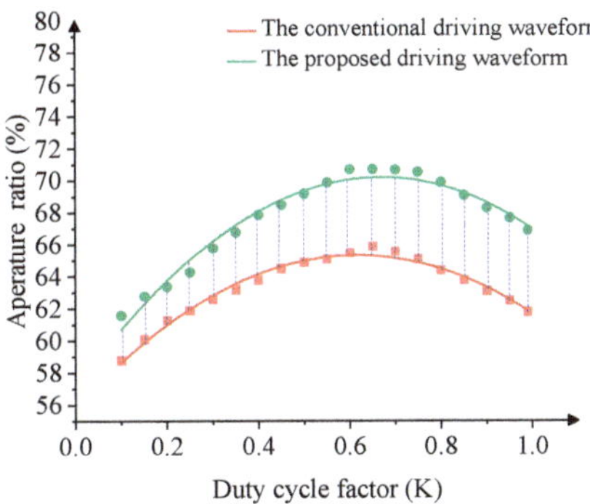

Figure 10. The relationship between the aperture ratio and the duty cycle in the driving waveform.

4.3. The Performance of Driving Waveforms

The aperture ratio and the response time are important factors in the driving waveform of EWDs. However, a slower slope of voltage rising speed can achieve a maximum aperture ratio in EWDs. In other words, the slower the voltage rising slope, the longer the response time required, which affects the display frame rate seriously.

As shown in Figure 11, the aperture ratio change trend of three driving waveforms are presented. Obviously, the shortest response time (8 ms) is achieved when the conventional driving waveform is applied, as shown in Figure 11a, but the ink is dispersed and the EWD aperture ratio is lowest. The response time becomes longer (65 ms) when a voltage slope is inserted into a driving waveform, as shown in Figure 11b; the ink is not dispersed and the aperture ratio of EWD increases significantly. In Figure 11c, the proposed driving waveform of this paper can achieve a shorter response time (15 ms). At the same time, the aperture ratio can also reach the maximum value (74%), which is about 8% higher than that of the conventional driving waveform. In Figure 11d, the proposed driving waveform can effectively control the shape of the ink, and a maximized aperture ratio is achieved. In order to obtain a clear comparison, parameter values among three kinds of driving waveforms are shown in Table 1.

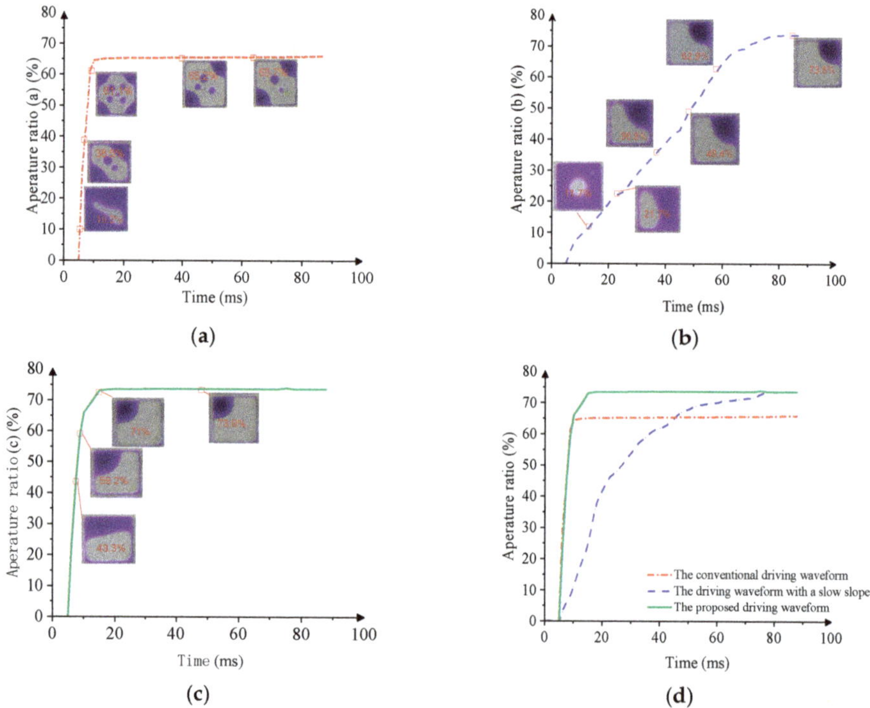

Figure 11. Change process of the EWD aperture ratio under driving waveforms. (**a**) Conventional driving waveform. (**b**) Driving waveform with a slow slope. (**c**) The proposed driving waveform in this paper. (**d**) Comparison of aperture ratio and response time among three driving waveforms.

Table 1. Comparison of parameter values among different driving waveforms.

Driving Waveform	Conventional Driving Waveform [6,7]	The Driving Waveform with a Slope [12]	Proposed Driving Waveform
Aperture ratio	65.7%	73.6%	73.9%
Response time	8 ms	65 ms	15 ms
Shape of ink			

5. Conclusions

In order to solve the problems of ink dispersion, hysteretic response, and low aperture ratio caused by the defect of the EWD dielectric layer, a driving waveform with a reverse electrode pulse and an optimized voltage slope was designed in this paper. A series of experiments were executed to test the driving waveform, specifically to include the pixel aperture ratio, frequency, and the duty cycle of the driving waveform. The results show that the aperture ratio can reach its largest value when the frequency of the driving waveform is about 60 Hz and duty cycle coefficient is 0.65. Compared with the conventional driving waveform, the aperture ratio increased by about 8%, and the ink steadily shrunk without dispersion. Hence, the display quality of EWDs was improved by optimizing the driving waveform.

Author Contributions: Z.Y. and L.W. designed this project. W.F. and Z.Y. carried out most of the experiments and data analysis. Y.L. and W.H. performed part of the experiments and helped with discussions during manuscript preparation. L.L. and Z.Z. contributed to the data analysis and correction. L.S., C.Z., and G.Z. gave suggestions on project management and conducted helpful discussions on the experimental results.

Funding: This research was funded by the Key Research Platforms and Research Projects in Universities and Colleges of Guangdong Provincial Department of Education (No. 2018KQNCX334), Zhongshan Innovative Research Team Program (No. 180809162197886), Guangdong government funding (No. 2014A010103024), Zhongshan Institute high-level talent scientific research startup fund project (No. 416YKQ04), Project for Innovation Team of Guangdong University (No. 2018KCXTD033), and the National Key R&D Program of China (No. 2018YFB0407100-02, No. 2016YFB0401502).

Conflicts of Interest: The authors declare no conflicts of interest.

References

1. Kim, D.Y.; Steckl, A.J. Electrowetting on paper for electronic paper display. *ACS Appl. Mater. Interfaces* **2010**, *2*, 3318–3323. [CrossRef] [PubMed]
2. Hayes, R.A.; Feenstra, B.J. Video-speed electronic paper based on electrowetting. *Nature* **2003**, *425*, 383–385. [CrossRef] [PubMed]
3. Edwards, A.M.J.; Brown, C.V.; Newton, M.I.; Mchale, G. Dielectrowetting: The Past, Present and Future. *Curr. Opin. Colloid* **2017**, *36*, 28–36. [CrossRef]
4. Beni, G.G.; Hackwood, S. Electro-wetting displays. *Appl. Phys. Lett.* **1981**, *38*, 207–209. [CrossRef]
5. Basu, A.S. Droplet morphometry and velocimetry (DMV): A video processing software for time-resolved, label-free tracking of droplet par-ameters. *Lab Chip* **2013**, *13*, 1892–1901. [CrossRef]
6. Luo, Z.J.; Zhang, W.N.; Liu, L.W.; Xie, S.T.; Zhou, G.F. Portable multi-gray scale video playing scheme for high-performance electrowetting displays. *J. Soc. Inf. Display* **2016**, *24*, 345–354. [CrossRef]
7. Yi, Z.C.; Shui, L.L.; Wang, L.; Jin, M.L.; Hayes, R.A.; Zhou, G.F. A novel driver for active matrix electrowetting displays. *Displays* **2015**, *37*, 86–93. [CrossRef]
8. Dijk, R.V.; Feenstra, B.J.; Hayes, R.A.; Camps, I.G.J.; Boom, R.G.H.; Wagemans, M.M.H.; Giraldo, A.; Heijden, R.V.D.; Los, R.; Feil, H. Gray Scales for Video Applications on Electrowetting Displays. *SID Symp. Dig. Tech. Pap.* **2006**, *37*, 1926–1929. [CrossRef]
9. Lu, Y.; Sur, A.; Liu, D.; Pascente, C.; Ruchhoeft, P. Dynamics of droplet motion induced by electrowetting. *Int. J. Heat Mass Trans.* **2017**, *106*, 920–931. [CrossRef]
10. Chiu, Y.H.; Liang, C.C.; Chen, Y.C.; Lee, W.Y.; Chen, H.Y.; Wu, S.H. Accurate-gray-level and quick-responsedriving methods for high-performance electrowetting displays. *J. Soc. Inf. Display* **2011**, *19*, 741–748. [CrossRef]
11. Xie, N.; Zhang, N.; Xu, R.Q. Effect of driving voltage polarity on dynamic response characteristics of electrowetting liquid lens. *Jap. J. Appl. Phys.* **2018**, *57*, 1–6.
12. Zhang, X.M.; Bai, P.F.; Hayes, R.; Shui, L.L.; Jin, M.L.; Tang, B.; Zhou, G.F. Novel Driving Methods for Manipulating Oil Motion in Electrofluidic Display Pixels. *J. Disp. Technol.* **2015**, *12*, 200–205. [CrossRef]
13. Roquescarmes, T.; Hayes, R.A.; Feenstra, B.J.; Schlangen, L. Liquid behavior inside a reflective display pixel based on electrowetting. *J. Appl. Phys.* **2004**, *95*, 4389–4396. [CrossRef]
14. Roquescarmes, T.; Hayes, R.A.; Schlangen, L.J.M. A physical model describing the electro-optic behavior of switchable optical elements based on electrowetting. *J. Appl. Phys.* **2004**, *96*, 6267–6271. [CrossRef]

15. Zhao, Q.; Tang, B.; Dong, B.Q.; Li, H.; Zhou, R.; Guo, Y.Y.; Dou, Y.Y.; Deng, Y.; Groenewold, J.; Alex, H.; et al. Electrowetting on dielectric: Experimental and model study of oil conductivity on rupture voltage. *J. Phys. D Appl. Phys.* **2018**, *51*, 195102. [CrossRef]
16. Moon, H.; Cho, S.K.; Garrell, R.L. Low voltage electrowetting-on-dielectric. *J. Appl. Phys.* **2002**, *92*, 4080–4087. [CrossRef]

 *micromachines*

Article

Different Regimes of Opto-fluidics for Biological Manipulation

John T. Winskas [1], Hao Wang [1], Arsenii Zhdanov [1], Surya Cheemalapati [1], Andrew Deonarine [2], Sandy Westerheide [2] and Anna Pyayt [1,*]

[1] Department of Chemical and Biomedical Engineering, University of South Florida, 4202 E. Fowler Ave. ENB118, Tampa, FL 33620, USA; John.Winsk@gmail.com (J.T.W.); wanghao@usf.edu (H.W.); arzhdanov@gmail.com (A.Z.); scheemalapati@usf.edu (S.C.)
[2] Department of Cell Biology, Microbiology and Molecular Biology, University of South Florida, 4202 E. Fowler Ave. ENB118, Tampa, FL 33620, USA; deonarine@mail.usf.edu (A.D.); Westerheide@usf.edu (S.W.)
* Correspondence: pyayt@usf.edu

Received: 23 October 2019; Accepted: 19 November 2019; Published: 21 November 2019

Abstract: Metallic structures can be used for the localized heating of fluid and the controlled generation of microfluidic currents. Carefully designed currents can move and trap small particles and cells. Here we demonstrate a new bi-metallic substrate that allows much more powerful micro-scale manipulation. We show that there are multiple regimes of opto-fluidic manipulation that can be controlled by an external laser power. While the lowest power does not affect even small objects, medium power can be used for efficiently capturing and trapping particles and cells. Finally, the high-power regime can be used for 3D levitation that, for the first time, has been demonstrated in this paper. Additionally, we demonstrate opto-fluidic manipulation for an extraordinarily dynamic range of masses extending eight orders of magnitude: from 80 fg nano-wires to 5.4 µg live worms.

Keywords: opto-fluidics; micro-manipulation; cells; microparticles

---

1. Introduction

Until now, to the best of our knowledge, the most common and reproducible methods of microscale manipulation have included mechanical and magnetic micro-manipulators [1,2] and various types of optical tweezers [3] including: holographic [4], plasmonic [5], antenna-based [6], and photonic crystal-based tweezers [7]. While these methods work well on smaller particles and nano-scale objects [8,9], they are limited in their application to biological specimens. They might induce severe heating of biological material [10] and phototoxicity [7]. More recently, magnetic levitation of cells has allowed for 3D manipulation of biological specimen and sensing capabilities that were unachievable using traditional 2D manipulation of cells, however it required the use of special magnetic fluids that are not compatible with many biological applications [2].

An alternative approach to cell manipulation is to use optofluidic tweezers (OFT) in capillaries [6] or thermally-induced current generation by absorption in amorphous silicon [11] using a light source focused through a microscope objective. These approaches are fundamentally different from the traditional optical and optoelectronic tweezers, and magnetic levitation, as they can attain much stronger forces [12,13]. Most recently, a thermo-plasmonic approach to OFT has appeared as the next generation of micro-manipulation techniques. In plasmon-assisted microfluidics, plasmonic heating is used to generate convective flow patterns [14–20]. Initially, simulations included the modeling of convection currents induced by photo-heating a single gold disk [14], which predicted very low fluid velocities in nm/s range. Later, it was demonstrated that localized heating of plasmonic structures could produce theoretically predicted toroid-shaped convection patterns [14–16]. Consequently, arrays of different plasmonic patterns were demonstrated [15], and this helped to somewhat increase the velocity. The currents could be generated by heating with different wavelengths of light in the near IR spectrum shining on a variety of plasmonic structures and rough metal films [14,17–20]. However, only small particles could be manipulated without causing significant heating. The main reason for that was a prior focus on optical properties of the whole system while ignoring its thermal properties. Later on, it was demonstrated that the thermal conductivity of the substrate plays an important role in the efficiency of the current generation. By adding continuous indium tin oxide (ITO) film to the substrate, the fluid velocity increased, and small particles were able to move with the speeds of up to 2 µm/s [21]. In addition to that, much faster flow can be demonstrated using thermo-plasmonic heating with the generation of a water vapor microbubble, but this requires a pretty high temperature of operation [22,23]. However, in all of these approaches to OFT the manipulated objects were exposed to significant amount of light radiation that could potentially damage cells or interfere with fluorescent imaging. Also, they were done using light coupled through a microscope objective, which made the whole setup inflexible in terms of independent observation and manipulation. Finally, they supported manipulation of small cells or particles on 2D surfaces, while we, for the first time, propose 3D manipulation and movement of much larger objects, truly pushing the limit of micromanipulation while using a very low controlling light power. This was accomplished because we were able to simultaneously optimize optical and thermal properties of our substrate to maximize the performance.

Our substrate contains a continuous bi-metal layer (Figure 1a), and this configuration has multiple advantages over traditional thermo-plasmonic structures. In comparison to small plasmonic patterns on transparent substrates, the continuous metal layer allows us to avoid the heating and phototoxicity that occurs due to direct light exposure of cells, while also providing the added advantage of, previously difficult, effective fluorescent imaging. The substrate is composed of a microscope glass slide with sputtered continuous bi-layer of chromium and gold. Light is absorbed by a thin layer of chromium, and the gold layer provides an interface that is biocompatible with cells and blocks the sample from light. Furthermore, the metallic bi-layer is optimized for maximum absorption by supporting a low-quality resonator, providing multiple passes of light through the chromium layer. This metallic bi-layer design can be optimized for various wavelengths by varying the thickness of both metals to optimize compound optical absorption and total thermal conductivity, allowing for the most effective micro-current generation. In this paper we demonstrate the safe and effective biological manipulation at two different wavelengths (532 and 808 nm) and show that the convective current generation can be efficiently supported with the help of the bi-metallic structure. Additionally, we demonstrate that opto-fluidic convection can be applied to several new biological functionalities including cell and particle capturing and levitation, fluid mixing, precise removal of biological material from surfaces, and cell/particle sorting. We also demonstrate that these phenomena can be optimized for ultra-efficient 3D manipulation of micro-scale biological objects, with seven orders of magnitude difference in mass, ranging from objects as tiny as silver nanowires to live organisms as large as an entire *Caenorhabditis elegans* worm.

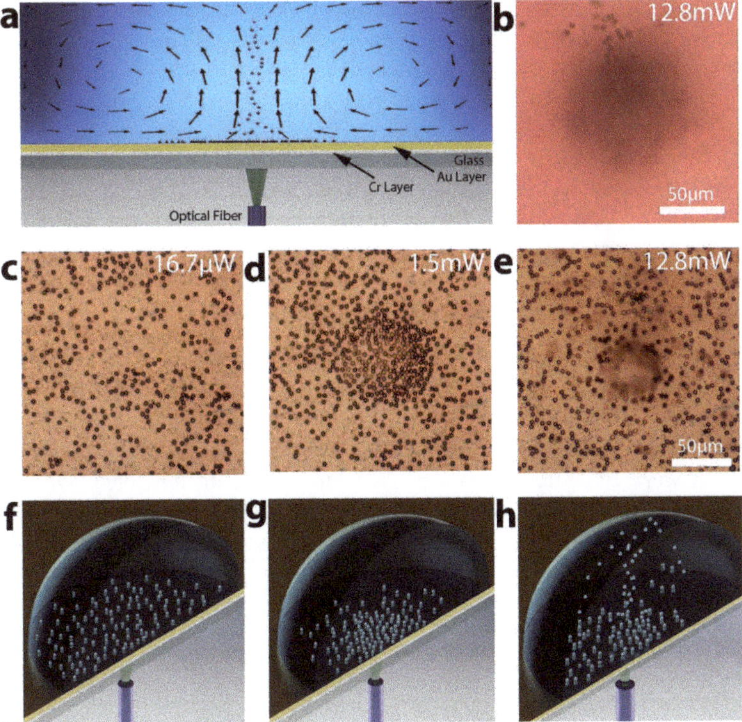

Figure 1. Explanation of opto-fluidic manipulation. (**a**) A 2-D model showing the trapping and levitation of particles using the convective vortices generated by the laser excitation. The arrows show fluid velocity vectors, and the thickness of arrows corresponds to the speed of movement. (**b**) Optical microscope image showing stream of 5 μm polystyrene beads levitated by opto-fluidics. (**c–h**) Optical microscope images and corresponding three-dimensional models of the three regimes of opto-fluidic particle manipulation; (**c,f**) low power—no controllable particle movement; (**d,g**) medium power—trapping regime; (**e,h**) high power—levitation regime. In (**e**) the spot in the center looks out of focus because in that area particles are projected upwards.

2. Methods

A schematic of the experimental setup is shown in Figure 1a. To manipulate micro-scale particles, we coupled a laser light into an optical fiber that had been cleaved on one end. The fiber was inserted into a fiber holder attached to a micromanipulator allowing XYZ fiber movement. After aligning the cleaved fiber tip underneath the microscope objective, we placed the substrate between the fiber and the objective. We then pipetted a drop of a media containing particles or cells on top of the gold surface. Next, the laser was turned on and light from the optical fiber was shining under the substrate, through the glass layer and chromium. By the time it reached the gold, it was partially attenuated by the chromium layer, it was then reflected by the gold layer and again propagated through the chromium experiencing additional attenuation. The gold layer protects the sample from the excitation light and enhances the attenuation and heat generated by the metallic bi-layer. Since the chromium thickness (5–20 nm) is significantly smaller than thickness of gold (200 nm), the thermal conductivity properties of the metallic bi-layer are dominated by gold. This results in a highly efficient local fluid heating. Continuous laser heating of the substrate locally warms the fluid and it continuously rises to the top of the droplet, this results in the formation of a consistent vortex shaped as a toroid (Figure 1a). We used the horizontal part of the current to push objects towards the center of the heated spot and the vertical part of the current to levitate particles with the stream of warm fluid upwards. This allows trapping relatively large objects, such as live nematodes and levitating a variety of smaller objects, such as red blood cells and micro-beads.

3. Results

Here we demonstrate that the intensity of the excitation light can be used to precisely control the fluid velocity and the forces applied to the microscopic objects. Our COMSOL simulations showed that the fluid velocity of the convective vortices is linearly proportional to the temperature of the heat source, while our experiments demonstrated that the temperature of the fluid is linearly proportional to the optical power. Thus, the fluid velocity is also linearly proportional to the optical power. This gave us the ability to easily and effectively adjust the fluid velocity and precisely control the size of the objects that we manipulated.

To study fluid velocity under different heating conditions, we used COMSOL multi-physics software that allows the simultaneous study of heat transfer and fluid dynamics. We simultaneously solve the Navier–Stokes equations and the conservation of energy equation to determine the velocity and the temperature fields. The micro-flow patterns formed by Rayleigh–Benard convection are analyzed.

Equations: Computation fluid dynamics is used to analyze the flow field. We are solving the continuity equation (Equation (1)), Navier–Stokes equations (Equations (2)–(4)), and conservation of energy equation (Equation (5)) for the corresponding initial and boundary conditions. Note that u, v, and w are components of the fluid velocity, ρ is density, T is the temperature, and k is thermal diffusivity. Conditions for Rayleigh–Benard convection are considered.

$$\frac{\partial u}{\partial x} + \frac{\partial v}{\partial y} + \frac{\partial w}{\partial z} = 0 \tag{1}$$

$$\frac{\partial u}{\partial t} + u\frac{\partial u}{\partial x} + v\frac{\partial u}{\partial y} + w\frac{\partial u}{\partial z} = -\frac{1}{\rho}\frac{\partial \delta}{\partial x} + f_x + v\left(\frac{\partial^2 u}{\partial x^2} + \frac{\partial^2 u}{\partial y^2} + \frac{\partial^2 u}{\partial z^2}\right) \tag{2}$$

$$\frac{\partial v}{\partial t} + u\frac{\partial v}{\partial x} + v\frac{\partial v}{\partial y} + w\frac{\partial v}{\partial z} = -\frac{1}{\rho}\frac{\partial \delta}{\partial y} + f_y + v\left(\frac{\partial^2 v}{\partial x^2} + \frac{\partial^2 v}{\partial y^2} + \frac{\partial^2 v}{\partial z^2}\right) \tag{3}$$

$$\frac{\partial w}{\partial t} + u\frac{\partial w}{\partial x} + v\frac{\partial w}{\partial y} + w\frac{\partial w}{\partial z} = -\frac{1}{\rho}\frac{\partial \delta}{\partial z} + f_z + v\left(\frac{\partial^2 w}{\partial x^2} + \frac{\partial^2 w}{\partial y^2} + \frac{\partial^2 w}{\partial z^2}\right) \tag{4}$$

$$\rho C_P\left(\frac{\partial T}{\partial t} + u\frac{\partial T}{\partial x} + v\frac{\partial T}{\partial y} + w\frac{\partial T}{\partial z}\right) = -k\left(\frac{\partial^2 T}{\partial x^2} + \frac{\partial^2 T}{\partial y^2} + \frac{\partial^2 T}{\partial z^2}\right) \tag{5}$$

The onset of natural convection is determined by the Rayleigh number Ra. For thermal convection due to heating from below, $\text{Ra} = \frac{\rho_0 \beta g \Delta T L^3}{\alpha \mu}$, where ρ_0 is the reference density, typically picked to be the average density of the medium, g is the local gravitational acceleration, β is the coefficient of thermal expansion, ΔT is the temperature difference across the medium, L is the characteristic length-scale of convection, α is a thermal diffusivity, and μ is the dynamic viscosity. Preliminary data based on COMSOL simulations show the velocity of the observed convection exhibits an approximately linear relationship with the laser power. Arrows representing velocity vectors simulated using COMSOL are shown in Figure 1a. We extracted the fluid pattern from the simulations and combined it with the illustration of the setup to simplify the understanding of the current structure. The simulations of the laser light-controlled currents were conducted for substrate temperatures ranging from 30 to 50 °C. In simulations at all temperatures, the current structure was very similar to the one shown in Figure 1a. The main difference was that all velocities were proportionally scaling up. While fluid velocity measured in the center of the droplet was ~1.2 µm/s at 30 °C, it reached 6.5 µm/s at 50 °C.

3.1. The Substrate Optimization, Fabrication, and Characterization

In contrast with previous studies, which used different types of patterned plasmonic structures and nanoparticles, we chose continuous metallic bi-layer. We found out from our COMSOL simulations that the thermal conductivity of the substrate greatly affects the efficiency of the current generation and its velocity. Figure 2a demonstrates that by heating three differently structured substrates (continuous gold layer, continuous chromium layer, and gold islands) to the same temperature, we observe a noticeable difference in the fluid velocity. Since gold has the highest thermal conductivity, the resulting convective current induced in the fluid is the most powerful. A continuous layer of chromium is less efficient due to its lower thermal conductivity and gold islands that do not form a continuous thermally-conductive layer were observed to be the least efficient. Therefore, the optimal structure should be continuous and have high thermal conductivity, preferably closer to gold.

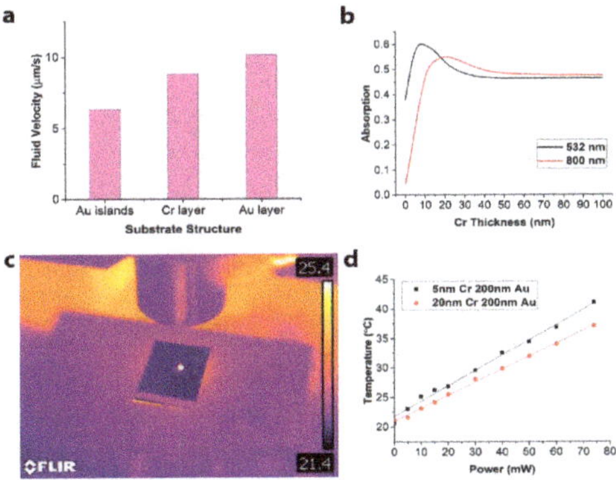

Figure 2. The substrate optimization. (**a**) COMSOL simulations were conducted to evaluate the most efficient substrate structure. (**b**) Finite difference time domain (FDTD) simulation results for substrate absorption optimization for 532 and 800 nm wavelengths varying chromium thickness from 0 to 100 nm while holding gold thickness constant at 200 nm. (**c**) FLIR IR camera imaging was used to measure the maximum temperature in a 1 µL drop of fluid. (**d**) Experimentally measured temperature increase while testing two substrates with 532 nm light with power ranging from 0 to 80 mW.

After determining the requirements for the thermal conductivity of the substrate, the next step was to take into consideration its optical and biocompatibility properties. A metallic bi-layer was chosen as a substrate because no single metal could simultaneously satisfy all our requirements: biocompatibility, high thermal conductivity, and high optical absorption for visible/IR wavelengths. Chromium was chosen for one of the layers for its high absorption for chosen wavelengths (532 and 800 nm) and gold was selected for its biocompatibility, thermal conductivity, and reflective properties. The individual metal thicknesses of the bi-layer were then studied to optimize overall performance. To optimize the optical absorption of the bi-layer we conducted finite difference time domain OptiFDTD simulations. The goal was to create a substrate that has opaque metallic bi-layer efficiently absorbing controlling light and isolating the sample from any light exposure. The gold layer thickness was chosen to be 200 nm for optimal reflection, while the thickness of the chromium layer was varied between 5 and 200 nm in the simulations (Figure 2b). Chromium and gold were both modeled as Lorentz–Drude dispersive materials. Optimization was conducted for wavelengths 532 and 800 nm because of wide availability of low-cost lasers with those wavelengths. In simulations the beams were linearly polarized, had a Gaussian profile, were emitting continuously, and were perpendicular to the substrate.

The absorption by bi-metallic substrate at both wavelengths with respect to the chromium thickness is plotted in Figure 2b. Since gold is highly reflective at both wavelengths, adding a layer of 200 nm of gold on top of chromium effectively doubles the propagation length of the light in the chromium layer. The absorption of the bi-layer is the highest for 5 nm of chromium at 532 nm and 20 nm of chromium at 800 nm. Interestingly, the optimized absorption by the layered structure was found to be 25%–30% higher than absorption by a single thick layer of Cr. This is the result of multiple reflections in the low-quality factor plasmonic resonator consisting of a metal bi-layer on a glass substrate.

Based on the results of the simulations, we fabricated substrates optimized for 532 and 808 nm lasers (5nm-Cr/200nm-Au and 20nm-Cr/200nm-Au, respectively), with the bi-layers deposited on glass slides by sputtering. After fabrication, the performance of the substrates was experimentally evaluated. All temperature measurements were conducted using a FLIR thermal camera. It was demonstrated that under local light exposure from an optical fiber, both substrates were able to efficiently absorb light and heat a 1 µL drop of water placed on top of the substrate (Figure 2c), reaching a maximum equilibrium temperature within 1–2 minutes after beginning of the experiment. As the power of the laser beam was increased from 0 to 80 mW, the temperature increased linearly from 20 to 40 °C (Figure 2d). This result, when combined with the linear relationship between the temperature and the fluid velocity vectors observed in COMSOL simulations, allowed us to infer a linear relationship between the fluid velocity in the generated convective vortex and the laser power.

Additionally, this result further demonstrated that an increase of 4 mW of the laser power results in a corresponding water temperature increase of 1 °C. Finally, it was experimentally confirmed that, similarly to the simulations shown in Figure 2b, the absorption at 532 nm is more efficient for chromium thickness of 5 nm comparing to 20 nm (Figure 2d), though both of them perform quite well.

3.2. Sample Manipulation Using Optimized Substrates

After optimizing the substrate, we demonstrated several new applications of 3D opto-fluidic manipulation to biological specimens including cell accumulation/trapping, projection, separation/filtering, followed by a viability study proving the safety of this approach to living cells. In the following experiments, we used human fibroblast cells suspended in a cell growth medium. The cells were treated with a 0.25% (w/v) Trypsin—0.53 mM ethylenediaminetetraacetic acid (EDTA) solution helping to temporarily keep the cells from bonding to the gold surface during experiments. We demonstrated the operation of the trapping regime at a controlling power of 15 mW using the 20 nm-Cr/200 nm-Au substrate optimized for 808 nm wavelength (Figure 3a–d). For this experiment a 5 µL drop of cell medium containing live fibroblast cells was placed on the substrate, and the optical fiber was aligned under an area of the substrate that initially contained no cells (Figure 3a). After 480 seconds, ten fibroblast cells were captured, and some of them from a distance greater than 250 µm (the field of view of the microscope).

To demonstrate the projection regime, we used a laser power of 47 mW applied to 5 nm-Cr/200 nm-Au substrate optimized for 532 nm wavelength. The sample was pipetted from a mixture of 1 µL of human blood and 1 mL of the medium containing fibroblast cells. The induced convection vortices were strong enough to levitate red blood cells vertically from the surface. After 260 seconds, we observed that both red blood cells and fibroblast cells were trapped in the center of the excitation area and there was a continuous stream of red blood cells projected upward (Figure 3e–h).

In the next set of experiments, it was observed that if we scan the optical fiber under the substrate (within the field of view), while operating in a high-power projection regime, we were able to successfully levitate and remove red blood cells from the surface without moving the fibroblast cells or trapping red blood cells (Figure 3i–l). The laser power used in these experiments was the same as in the previous experiment (47 mW). When optical fiber moves, the toroid current pattern does not have enough time to form, and the vertical water movement simply projects the red blood cells off the surface, like a miniature "pressure washing" system. These tree regimes—trapping, levitation, and pressure washing can be in parallel and differently applied to different types of particles with applications in trapping, mass-based filtrations/separations, measurements of masses of individual cells, and sensing applications.

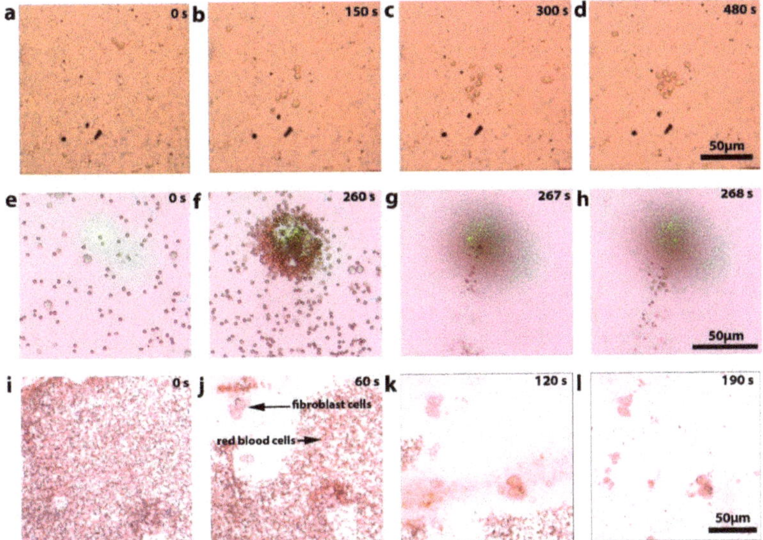

Figure 3. Experimental demonstration of several opto-fluidic manipulation regimes. (**a–d**) Trapping of fibroblast cells during 480 second period. (**e–h**) Trapping of heavy fibroblast cells simultaneously with the projection of lighter red blood cells. (**e,f**) Microscope imaging plane of the substrate, (**g,h**) imaging above the substrate demonstrates stream of red blood cells being levitated. (**i–l**) Targeted "pressure washing" removing light red blood cells from the surface leaving in place heavy fibroblast cells. Optical fiber scans under the whole portion of the substrate visible in the field of view selectively removing one type of the cells from the surface.

Finally, the following experiments were conducted to demonstrate the extraordinary dynamic range of masses that could be manipulated using this technology, along with showing that this approach can also be used to build complex, multi-layer structures (Figure 4). First, we trapped live *C. elegans* worms using a 35 mW laser power (Figure 4a–d). The worms were still active after the manipulation, showing that the temperature increase did not damage them. The live worms are the largest reported living organisms manipulated using light. They have an average length of 1 mm and width of 50 μm. In contrast to the macroscopic worms, we show that the same technique can be used to manipulate objects as small as silver nanowires which have masses seven orders of magnitude smaller than that of worms (Figure 4m,n). Using our opto-fluidic manipulation we were not only able to capture silver nanowires from the solution but demonstrated the ability to levitate them and build multi-layer structures. In future this can be used for building more complex 3D nano-scale structures without the use mechanical manipulators. Additionally, we also demonstrated that our approach can be used for selective size-based trapping and assembly (Figure 4e–l). We were able to separate 20 μm polystyrene micro-beads from their mixture with 5 μm beads by trapping larger particles and projecting away smaller ones. The 20 μm beads assemble in multi-layer structure that can be potentially used for building photonic crystals and other complex ordered assemblies.

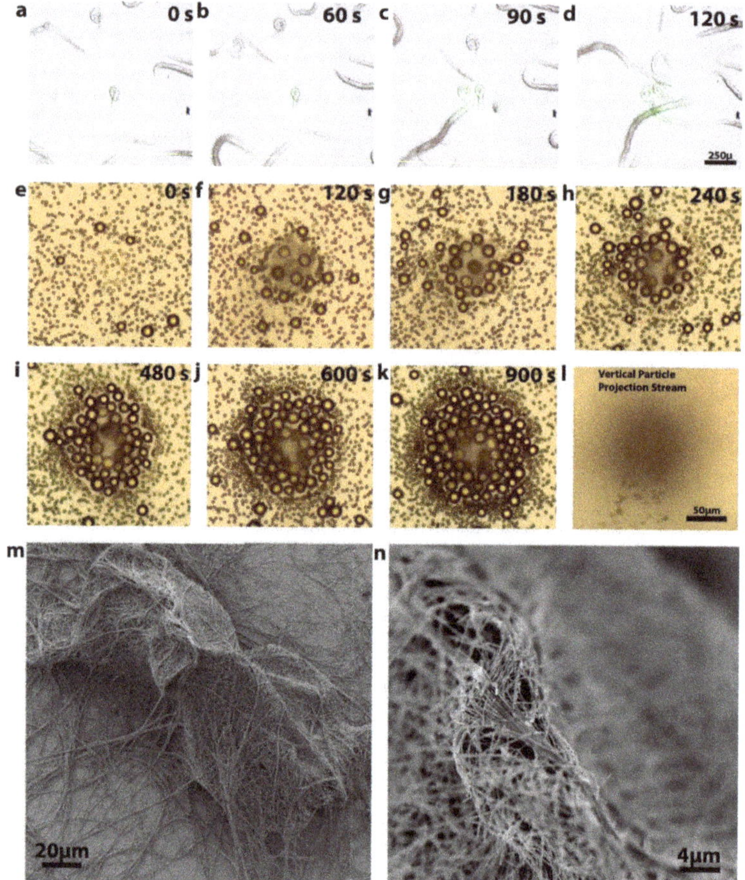

Figure 4. Demonstration of the extraordinary dynamic range of opt-fluidic manipulation and application to the assembly of complex structures. (**a–d**) Capturing live worms using 40 mW power. Worm length varied from 50 to 200 um. (**e–l**) Selective capturing of large particles and their multi-layer assembly. (**m,n**) A multi-layer silver nanowire structure built using trapping and levitation.

3.3. Cell Viability

To test the biocompatibility of the technology, a series of cell viability tests were conducted for different power regimes used in cell manipulation experiments (Figure 5). To ensure all cells used were studied under the same conditions, we integrated the substrate with a PDMS chip containing multiple identical wells. Each well was used to test cell viability after manipulation using one of the regimes, in addition to wells used for control experiments without any manipulations. The fibroblast cells were manipulated in low-power regime (3.7 mW), medium-power trapping regime (38 mW), and the high-power projection regime (78 mW) (Figure 5a1,a2,b1,b2,c1,c2). The cells in each well were manipulated for 150 seconds, after which, the whole chip was returned into an incubator, where the cells were cultured for 24 hours. During that time, the cells bonded to the gold surface and started growing. Bright-field microscopy demonstrated that the cells looked healthy and had an expected morphology (Figure 5a3,b3,c3). We then used a live/dead cell imaging kit to stain live cells with green-fluorescence-emitting dye, and dead ones with red-emitting dye. It was confirmed using fluorescence imaging that the cells appeared healthy and emitted a green fluorescence signal

(Figure 5a4,b4,c4). After a detailed examination, on average, less than one dead cell was observed per well, matching the results from the control well that was not exposed to any manipulations. This result confirmed that all regimes of opto-fluidic manipulation were biocompatible and did not decrease cell viability.

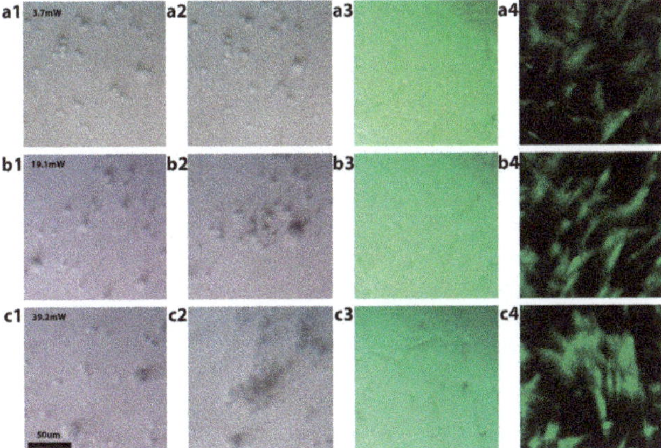

Figure 5. Demonstration of fibroblast cell viability after opto-fluidic manipulation. (**a1,a2**) Before and after low-power manipulation at 3.7 mW. (**b1,b2**) Before and after medium-power trapping at 38 mW. (**c1,c2**) Before and after high-power projection regime at 78 mW. (**a3,b3,c3**) Bright-field and (**a4,b4,c4**) fluorescent images of cells after 24 hours of incubation.

4. Conclusions

To summarize, we have demonstrated for the first time several new regimes of opto-fluidic manipulation. We showed the trapping of objects with the extraordinary seven orders of magnitude dynamic range of masses, from microscopic nano-wires to macroscopic live *C. elegans* worms. Furthermore, our studies have shown that we not only captured, but levitated, filtrated, and "pressure washed" a variety of objects. On top of this, we have observed, through cell viability testing, that this technology can be safely used, not only for capturing but also the levitation and separation of cells and other biological organisms. We also show that this method can be customized for different wavelengths of light. This opens the door for consistent, reproducible, and easily fabricated devices that can be implemented across a wide range of fields that require gentle and powerful biomanipulation of objects and cells, and small live organisms. We suggest that some of the potential applications might include filtering different types of cells, separation of bacteria or circulating tumor cells from biological samples. Additionally, capturing worms using our approach is very attractive, since all traditional mechanical ways to move the worms around require use of sharp metal tips and other objects that damage worms, while optical tweezers would require extremely high power that would kill them. At the same time, our approach allows gentle movement of live worms to a needed location for further imaging and analysis.

Author Contributions: A.P. lead the project, conceived and designed the experiments and simulations; J.T.W. and H.W. performed the experiments; S.C. conducted simulations, A.D. and S.W. helped with C. elegans experiments, A.Z. created visualizations, all authors contributed to manuscript writing.

Funding: Dr. Anna Pyayt's Lab was supported by NSF Grant # 1701081.

Conflicts of Interest: The authors declare no conflict of interest.

References

1. Lee, W.R.; Oh, K.T.; Park, S.Y.; Yoo, N.Y.; Ahn, Y.S.; Lee, D.H.; Youn, Y.S.; Lee, D.K.; Cha, K.H.; Lee, E.S. Magnetic levitating polymeric nano/microparticular substrates for three-dimensional tumor cell culture. *Colloids Surf. B Biointerf.* **2011**, *85*, 379–384. [CrossRef]
2. Durmus, N.G.; Tekin, H.C.; Guven, S.; Sridhar, K.; Yildiz, A.A.; Calibasi, G.; Ghiran, I.; Davis, R.W.; Steinmetz, L.M.; Demirci, U. Magnetic levitation of single cells. *Proc. Natl. Acad. Sci. USA* **2015**, *112*, E3661–E3668. [CrossRef]
3. Thoumine, O.; Ott, A.; Cardoso, O.; Meister, J.J. Microplates: a new tool for manipulation and mechanical perturbation of individual cells. *J. Biochem. Biophys. Methods* **1999**, *39*, 47–62. [CrossRef]
4. Korda, P.T.; Taylor, M.B.; Grier, D.G. Kinetically locked-in colloidal transport in an array of optical tweezers. *Phys. Rev. Lett.* **2002**, *89*, 128301. [CrossRef]
5. Schuller, J.A.; Barnard, E.S.; Cai, W.; Jun, Y.C.; White, J.S.; Brongersma, M.L. Plasmonics for extreme light concentration and manipulation. *Nat. Mater.* **2010**, *9*, 193. [CrossRef]
6. Kurup, G.K.; Basu, A.S. Rolling, aligning, and trapping droplets on a laser beam using marangoni optofluidic tweezers. In Proceedings of the 16th International Conference Solid-State Sensors, Actuators and Microsystems (TRANSDUCERS), Beijing, China, 5–9 June 2011; pp. 266–269.
7. Khan, I.; Tang, E.; Arany, P. Molecular pathway of near-infrared laser phototoxicity involves ATF-4 orchestrated ER stress. *Sci. Rep.* **2015**, *5*, 10581. [CrossRef]
8. Berthelot, J.; Aćimović, S.S.; Juan, M.L.; Kreuzer, M.P.; Renger, J.; Quidant, R. Three-dimensional manipulation with scanning near-field optical nanotweezers. *Nat. Nanotechnol.* **2014**, *9*, 295–299. [CrossRef]
9. Mandal, S.; Serey, X.; Erickson, D. Nanomanipulation using silicon photonic crystal resonators. *Nano Lett.* **2009**, *10*, 99–104. [CrossRef]
10. Liu, Y.; Cheng, D.K.; Sonek, G.J.; Berns, M.W.; Chapman, C.F.; Tromberg, B.J. Evidence for localized cell heating induced by infrared optical tweezers. *Biophys. J.* **1995**, *68*, 2137–2144. [CrossRef]
11. Chiou, P.Y.; Ohta, A.T.; Wu, M.C. Massively parallel manipulation of single cells and microparticles using optical images. *Nature* **2005**, *436*, 370. [CrossRef]
12. Kurup, G.K.; Basu, A.S. Hydrodynamic particle concentration inside a microfluidic plug. In Proceedings of the 14th International Conference on Miniaturized Systems for Chemistry and Life Sciences (MicroTAS), Groningen, The Netherlands, 3–7 Octeober 2010; pp. 740–742.
13. Trivedi, V.; Doshi, A.; Kurup, G.K.; Ereifej, E.; Vandevord, P.J.; Basu, A.S. A modular approach for the generation, storage, mixing, and detection of droplet libraries for high throughput screening. *Lab Chip* **2010**, *10*, 2433–2442. [CrossRef]
14. Miao, X.; Wilson, B.K.; Lin, L.Y. Localized surface plasmon assisted microfluidic mixing. *Appl. Phys. Lett.* **2008**, *92*, 124108. [CrossRef]
15. Donner, J.S.; Baffou, G.; McCluskey, D.; Quidant, R. Plasmon-assisted optofluidics. *ACS Nano* **2011**, *5*, 5457–5462. [CrossRef]
16. Chen, J.; Kang, Z.; Kong, S.K.; Ho, H.P. Plasmonic random nanostructures on fiber tip for trapping live cells and colloidal particles. *Opt. Lett.* **2015**, *40*, 3926–3929. [CrossRef]
17. Kang, Z.; Chen, J.; Wu, S.Y.; Chen, K.; Kong, S.K.; Yong, K.T.; Ho, H.P. Trapping and assembling of particles and live cells on large-scale random gold nano-island substrates. *Sci. Rep.* **2015**, *5*, 9978. [CrossRef]
18. Lozan, O.; Perrin, M.; Ea-Kim, B.; Rampnoux, J.M.; Dilhaire, S.; Lalanne, P. Anomalous light absorption around subwavelength apertures in metal films. *Phys. Rev. Lett.* **2014**, *112*, 193903. [CrossRef]
19. Meier, M.; Wokaun, A.; Liao, P.F. Enhanced fields on rough surfaces: dipolar interactions among particles of sizes exceeding the Rayleigh limit. *JOSA B* **1985**, *2*, 931–949. [CrossRef]
20. Norman, T.J.; Grant, C.D.; Magana, D.; Zhang, J.Z.; Liu, J.; Cao, D.; Bridges, F.; Van Buuren, A. Near infrared optical absorption of gold nanoparticle aggregates. *J. Phys. Chem. B* **2002**, *106*, 7005–7012. [CrossRef]
21. Jeong, Y.G.; Lee, J.S.; Shim, J.K.; Hur, W. A scaffold-free surface culture of B16F10 murine melanoma cells based on magnetic levitation. *Cytotechnology* **2016**, *68*, 2323–2334. [CrossRef]

22. Namura, K.; Imafuku, S.; Kumar, S.; Nakajima, K.; Sakakura, M.; Suzuki, M. Direction control of quasi-stokeslet induced by thermoplasmonic heating of a water vapor microbubble. *Sci. Rep.* **2019**, *9*, 4770. [CrossRef]
23. Flores-Flores, E.; Torres-Hurtado, S.A.; Páez, R.; Ruiz, U.; Beltrán-Pérez, G.; Neale, S.L.; Ramirez-San-Juan, J.C.; Ramos-García, R. Trapping and manipulation of microparticles using laser-induced convection currents and photophoresis. *Biomed. Opt. Express* **2015**, *6*, 4079–4087. [CrossRef]

MDPI
St. Alban-Anlage 66
4052 Basel
Switzerland
Tel. +41 61 683 77 34
Fax +41 61 302 89 18
www.mdpi.com

Micromachines Editorial Office
E-mail: micromachines@mdpi.com
www.mdpi.com/journal/micromachines